A ZOOLOGIST LOOKS AT HUMANKIND

A ZOOLOGIST LOOKS AT HUMANKIND

Adolf Portmann

TRANSLATED BY JUDITH SCHAEFER

COLUMBIA UNIVERSITY PRESS NEW YORK

COLUMBIA UNIVERSITY PRESS
NEW YORK OXFORD

Library of Congress Cataloging-in-Publication Data

Portmann, Adolf, 1897–
[Biologische Fragmente zu einer Lehre vom Menschen. English]
A zoologist looks at humankind / Adolf Portmann;
translated by Judith Schaefer.
p. cm.
Translation of: Biologische Fragmente zu einer Lehre vom Menschen.
Includes bibliographical references.
ISBN 0-231-06194-3
1. Man—Origin. 2. Human evolution—Philosophy. 3. Biology—Philosophy. 4. Human growth. I. Title.
GN281.P6713 1990
573.2—dc20 89-22203
CIP

Casebound editions of Columbia University Press books are Smyth-sewn and printed on permanent and durable acid-free paper

Printed in the United States of America
c 10 9 8 7 6 5 4 3 2 1

Contents

Preface

THE STUDIES presented in these "fragments" were carried out in the years following 1939 and appeared for the first time in 1944. They have slowly acquired their readers. In 1951 a second edition appeared. In 1956, the brief work was reissued in soft cover as part of the *Rowolts deutscher Encyklopädie* series (volume 20), under the title *Zoologie und das neue Bild des Menschen,* which has meant that, in the last ten years, these preliminary studies for a more comprehensive anthropology have received wider distribution. At the time of reissue, the 1944 text was altered in a few places by the addition of essential supplementary material.

Since then, biological research has produced much that is new, to which many of my own studies and those of my colleagues have contributed. Consequently, the plan for a new edition called for a thorough review and raised the question of whether the book should be expanded. Was that not the moment to refashion *Biologische Fragmente* into a larger work, one for which a considerable amount of preliminary study was either already available or in process?

But there was something special about the little book of 1944. It represented a step forward for anthropology. It proved to be the introduction to a special problem, that of taking a comprehensive look at research that had often been disastrously isolated within the narrow confines of academic departments. Another factor in the decision was that, although new studies had clarified or modified many details, they indeed validated the fundamental concept upon which the 1944 manuscript was based. This caused us to reissue *Biologische Fragmente* and give the planned larger work more time to mature.

The form has been retained as far as possible. The factual information has, nontheless, been reviewed and expanded. This is particularly true for the data on the development of prosimians and, further, for the description of cerebralization. The discussion of accelerated development during our youthful stage was reworked and given a section of its own, which follows extensively an academic lecture I gave in Basel in 1966. The chapter on aging has also undergone some changes. I thank my colleagues at the Zoological Institute and the local zoo for the opportunity to present important zoological facts through illustrations that originated right here in Basel; in the last ten years, the zoo has made particularly significant contributions to the developmental history of, and reproduction in, large mammals.

In their original form, these "fragments" helped to form a multifaceted view of humanness. They pointed out paths biology is following as it strives for a comprehensive understanding of our form of existence, changing, as it does so, into a new branch of anthropology. Moving beyond the barriers of the prescientific divisions of body, soul, and mind, we see that the task is one.

The book came into being during years of the greatest uncertainty as to our fate, a time that demanded the deepest, most serious contemplation of the special nature of the human form of life. The moment in which the fragments appear again is not less somber. I hope this little book in its new form continues to fulfill its original mission.

Adolf Portmann
Basel, October 1968

A ZOOLOGIST LOOKS AT HUMANKIND

Introduction

AS THE life sciences have increasingly elevated evolutionary thought to be their guiding principle, biological facts and theories have come to make a decisive contribution to the various views of man held today.

In the second half of the last century, the importance of biological research grew considerably. Fostered by the life-science orientation of psychology and ethnology, biological research contributed to that course of events which, taken as a whole, has been sometimes extolled as progress in the unmasking of fallacious ideals, sometimes condemned as the infamous degradation of all spiritual values: namely, the biological explanation of human ethical and religious ideals as being an ideological superstructure, a disguise for the baser, more animalistic drives. Biology began to assume an importance that procured for it a central role in many very recent political concepts and theories. This role, however, has long been in preparation.

New Concepts of the Organism

FOR SOME generations now, the life sciences have more or less consciously helped to shape a completely new concept of the organism. Thus, about three hundred years ago, microscopic observation of the development of the germ cell introduced into the general awareness the fact that each individual was formed from a "germ," the structure of which was scarcely visible even through the microscope. The eggs from the most widely differing animals could look so much alike that one could not discern, simply by looking through the microscope, the different kinds of creatures they would give rise to. A new field of research came into being. That semitransparent, living substance, which seemed so simple, had to contain everything necessary to form a richly complex body—a coral, a bird, or a human. Today, using the methods of physics and chemistry, and aided by bacteriology and genetics, we search for those invisible structures within the egg; investigation of the submicroscopic structures concealed within an optical "void" has become one of the most important areas of recent biological research.

Scientific probing of the processes that take place within the protoplasm has made itself felt far beyond the actual area of research. This kind of investigation has reinforced the tendency throughout Western thought, already influenced by the rising flood of discoveries in natural history, to move away from the ancient idea of Creation: away from an idea of "created beings" or "creatures"—forms brought into being by a creative power—toward the notion of organisms that create themselves, of the germ that "develops on its own." Until late in the eighteenth century, this development was thought to consist of the literal "unfolding" of an invisible, preformed complexity, a concept that was basically still linked, however imperceptibly, to the idea of an incomprehensible clockwork mechanism created but once. Behind this concept stood always the Creator, pale and distant, perhaps, but still so powerful that questions of origin were long avoided by biology. The nineteenth century and fresh developments brought radical change: the new concept was a truly neogenetic, visibly complex form arising from an invisibly structured germ cell with particular powers of structural formation. Indeed, in recent decades, the study of developmental phsyiology has successfully begun to delimit more precisely what in the formation of a single being—in ontogeny—is actually a newly created form, and what is development from invisible structures contained within the ovum. Scientists agree that all capacity for development of form and structure is latent in minute quantities of living matter, which, in the human germ cell, for example, can be as small as 0.2 mm in diameter.

The intellectual transformation that will arise from those discoveries has scarcely begun. The first generalizations drawn from the knowledge of the self-differentiation of the germ will yet set off much reaction; and many new advances will have to be made before these insights become fixed images, as deeply rooted in human consciousness as those ancient concepts of the Creation that still hold sway over us.

The contrast between the void seen through the microscope and the wealth that we must assume lies within the germ tempts us to ascribe more and more power to its formative capacities, which must, indeed, be granted. The rich get richer, as the saying goes.

Observation justifies, even compels, investing the optically simple germ cell with the capabilities of forming an animal—and widely differing variants of that animal! On the other hand, it is going too far also to attribute to this germ cell the possibility of forming such novelties as would cause entirely new forms to appear. This is the danger inherent in the new concept of the organism. It inclines us to expand the actually observed capacity for morphological development to explain all morphological relationships of organisms as coming from one geneological context. And it is not the extrapolation as such that is the danger, for the mind is always making bold sallies beyond observable facts into the unseen. The danger arises when such intellectual flights are taken all too easily as truth and, as such, influence broader aspects of human life.

We know of large numbers of genetically fixed, suddenly appearing variants of plant and animal species. They are called mutations, and it should go without saying that this word applies only to mutations actually observed. No one will underestimate the importance of observed mutations and of the facts established based on observed mutations. But today, this term has been extended by analogy to apply not only to what has been observed but also to countless supposed mutations, which, in ever new transformational steps eventually lead from fish to amphibian, from reptile to bird. Such conclusions drawn from experimental research in mutation exceed anything that can be established empirically. Mainly, however, the term itself, which sounds so axiomatic, conceals the fact that only too often today this magic word *mutation* allows us to profess a knowledge of events that no one can know of with certainty.

The same casual attitude associated with the word *mutation* is also reflected in the superficial use of the word *nature,* which many people today, in a very tangible sense, invest with all the meaning once ascribed to the agency of a creative force. Instead of signifying the exceptional appeal to the unnamable, uttered in full awareness of the gravity of the word, *nature* today has become a facile means of verbal explanation, just good enough to let us know that everything is going according to plan—and to allow the illusion to persist that we know the deeper ground of being, which, however, surrounds us as ever, unknown and incommunicative.

Knowledge of the development of the germ cell has contributed to a fundamental transformation of our idea of the organism—the impression of self-development, of "self-creation" by the organism, has come to have dominion over our thought. We do not want to forget, however, that this only states the great riddle of evolution and does not solve it, and that investing the germ cell with further potential for the creation of forms—for the creation of new types, for example—far exceeds the limits of scientific assertion and draws its explanatory value from the fallow religious energies of the present day.

The new concept of the organism is a powerful force in the vital sentiments dominating our time. The facts of embryology have granted the concept far more power and credit than it can support—it lives beyond its means. This credit contributes to the conviction that the mystery of the living organism is fully accessible to investigation and research. The thought that in life-forms there exist not only earthly phenomena but cosmic phenomena, too, is supplanted by the idea that the entire riddle of all morphological relationships lies before us, still locked within the invisible structure of the protoplasm, to be sure, but accessible to research nonetheless! No one can underestimate the powerful stimulus that such an idea can occasionally provide. If, however, the research findings prompted by this idea lead to a broader view of reality, the thought must also remain alive in experimental biologists that any insight into the onset of germ-cell development says nothing about the problem of the actual origin, and that research into the germ cell can only proceed from the already existing complex structure of the ovum. How much will the probing of the submicroscopic structures of the protoplasm of the final given—the species-specific protoplasm—ultimately reveal?

Extrapolating the observation of self-creating organisms to the idea of an entire animal phylum arising from self-differentiating plasma has also powerfully advanced the thought that we could solve the riddle of humankind by studying animals—that the higher form is explained by the simpler one, that the form and substance of the human being could be fathomed by biological research on primates. Considerable impetus for intensive research on primates proceeds from this basic idea. We probably do not need to say anything about the manifold significance of studies on great apes and other related mammalian forms, but we should have no illusions as to the special nature of this significance. For it is not that primate studies have provided information on brain function or sexual cycles and early stages of development because primate organization presents these phenomena somewhat more authentically, typically, or perhaps somehow more conclusively than does the human body. No, all these studies are important primarily because we can perform on the bodies of other primates operations that we may not perform on humans. All valid results of primate research belong to the group of facts thus obtained, whereas the theoretical trains of thought that have been triggered by theories of evolution have often obstructed more true insight into the human race than they have revealed.

The idea that the higher form derives from the lower leads to error. The essence of the higher form cannot be understood through the condition of the lower form even if it is extremely likely that the higher form descended from the lower one. For example, I elucidate the peculiarities of bird flight much more by using the flight organs of bats for comparison than by studying the forelimbs of reptiles, from which both flight structures derive. However conclusive and important the evidence of morphological similarity between the bird wing and the reptile leg may be, the most essential quality that distinguishes the bird wing is not illuminated by the comparison.

As we thus limit the possibilities for explanation that the study of the lower stages of form offers for the understanding of the higher forms of life, we must, then, also emphasize—more than is usually done—what we owe to the knowledge of our own inner life for the understanding of all animal existence. There is also a continuous stream of interpretation flowing from our own experience into our biological work with animals, a stream that can only come from that special wellspring of our own experience. This subjectivity should not be perfunctorily deemed suspect for being all too human, but, rather, should be made use of in a meaningful way. The vision of life looking down from above, from the point of view of the human being is a necessary complement to the attempt at building from beneath, to proceeding from the simplest forms. This vantage point is indispensable in the overcoming of an interpretation that appeared with the untenable claim of offering an ultimate concept of the origin of organic forms.

Conscious and Unconscious Life

OF THE changes wrought by the weight of biological research in the general thinking, changes that have had far-reaching consequences for the concept of the human being, the devaluation of consciousness is not the least. In a time when the research methods of natural history strive so hard to express all the properties of an organism in size and number, the insight must be particularly apparent that the realm of conscious experience within the spatial and temporal extent of the profusion of earthly life is infinitely small—just a dot, really—and that this consciousness occurs as if banished to tiny islands in a sea of unconsciously lived life.

In addition, the weight and significance of this unconscious existence has been powerfully increased by growing insight into the complicated controls of all life processes exercised by the astonishing, incredible power of the protoplasm. Furthermore, knowledge of the manifold physical restrictions imposed on conscious states, discoveries of the influence of hormones and instinctual drives on these processes—all contribute to the devaluation of conscious existence. Not only does the life of the conscious being seem insular amid a mighty heave and billow of vegetative existence, but within the human being, too, the conscious appears to be embedded in an unperceived force for ordered existence: watchful, all-enveloping, and preordained.

With this knowledge comes the unmasking of so many of the creations of conscious intellectual and spiritual life, revealing them to be the results of dark instincts, ideological superstructures, tools for those in power who secretly use such consciousness as an instrument to further their own ends. What a change from the time when our body was regarded as the lowly vessel of the spirit! Increasingly, the life of the mind, once regarded as being so free and independent, appears to be the product of unknown powers, which, however, do not interfere with and control our existence from some distant heaven, but rather operate in a predetermined way from within the organic structure of our species. The works of the mind appear as the fruit of an unconscious that is powerfully influenced by substances in the blood and controlled by genetic constraints, the innate structure of the nervous system, and predetermined ways of thinking and perceiving. Taken all together, these limiting factors are often understood as the common genetic inheritance of constitutional types of races.

Small wonder that, as a consequence of such views—when characteristics common to larger human groups, traits by means of which the individual shows himself to be the member of a group, are given precedence—attention turns more and more away from the uniqueness of the individual. Consideration is given to what the commonality creates, which, in the extreme, is defined as what is valuable. Anything that goes against the group concept is strongly condemned as "deviation" from the type. In science, this trend toward establishing group-typical characteristics leads to research into types, which, if carried on in an objective, considered way, could be a significant instrument of knowledge. However, the danger of overrating its concepts too hastily, of sloganlike misuse of unverified, partial results has not been avoided, with the result that this *Konstitutionsforschung** purposefully or unwittingly contributes to the devaluation of the sphere of independent, intellectual decision-making in favor of a higher appreciation for all modes of mental activity that may be construed as group characters.

Conceiving of the mind as an organ with obscure vital powers has shaken values heretofore in force so seriously that, in the end, the inherited group characteristics emerged as about the only factors that were somewhat reliable and which necessarily came to be the mainstay of the devaluation process. Ultimately, many of these

*Research into human constitutions conducted by German psychiatrist Ernst Kretschmer (1888–1964); in an attempt to correlate body build with personality, character with mental illness, he established three types: the asthenic (tall, thin); the pycnic (thickset, squat); and the athletic (muscular).——Trans.

group characteristics provided standards for life; as "race" or "blood" they were elevated to the status of powers and increasingly came to be revered in the manner once reserved for supernatural forces.

The heavy emphasis on the unconscious part of our existence is one of the reactions against the earlier appraisal of the callous, unfettered workings of conscious intellectual activity. Disillusionment over the repeated misuse of the fruits of intellectual activity as a weapon in the brutal struggle for existence; disappointment that many systems of thought were inadequate to cope with life's problems; the stark contrast between scientific progress and daily life: all of these complex factors interacted to contribute to a disdain for conscious existence, to its debasement. This last found a powerful foil in the great beauty of the free, unconscious life of nature, where man has no influence; such wild, unbridled grandeur intensified the feeling of intellectual impotence and emphasized the painful contradiction.

The thinking outlined above contributed to the tragic events that led to the Second World War, and the consequences of those terrible years are certainly evident to all. However, the moment the pervasive disillusion with intellectuality is clearly seen as the source of the devaluation the worst is over, and we are encouraged once more to seek the elusive connections between unconscious and conscious life without passing judgment in advance on one or the other aspect. Future research will take into consideration at every step all the implications of the fact of mankind; it will also bear in mind, however, that the very attempts to unmask conscious intellectual activity, thereby revealing its supposed degraded nature, have been carried out by the intellect itself, and that, therefore, this curious attempt at self-mutilation is possible only through a deep, albeit unacknowledged, complete recognition of the tool with which it has been undertaken.

The change in the evaluation of the relationship between the conscious and the unconscious aspects of the human being will lead to an objective concept only when the shakiness of one of the most influential underlying principles of contemporary biological thought and deed is recognized: when the biologically grounded theory of evolution is demoted from the rank of self-evident truth to the simpler category of important, fruitful biological theory.

The Two Concepts of Evolution

OF THE many ideas that have arisen within the field of biology, none has had a broader, more varied effect on contemporary life than the concept of organic evolution. That is why, moreover, no other concept has become more significant for our concept of the human being.

In the life of our time, the theory of evolution has had the force of irrevocable truth, one wrested by research from mysterious obscurity. And yet, how completely different is the meaning of evolutionary theory if, when seeking to understand it, one applies the detachment a scientist must use (at least now and then) in confronting the results of his research. Seen within the context of the ultimate goal of research, which is to find the nature of reality, evolutionary theory is a bold move to explain enormous masses of facts. However, in all its particulars it remains the object of stringent criticism and discussion, and in its ultimate ramifications it leads at every turn beyond what is comprehensible scientifically to the realm of faith. It is, then, also from the contemporary, deeply felt need to believe in something that this bold attempt acquires its special force: Many people are convinced that this attempt at explanation contains within it the basis for a view of the world and of mankind that is consistent with our knowledge, something the old forms of belief are no longer able to provide.

It is hardly necessary to say how fruitful the concept of evolution has been, and will continue to be, for biology. As long as we are clear about the theory's place as an attempt to explain the somatic relationships—so difficult to understand—of plants and of animals, we will continue to be able to use this working principle effectively to forge ahead in the struggle with concealed reality. No other idea is currently in a position to provide a scientific explanation for the succession of animal and plant forms so impressively demonstrated by paleontology. The concept is also a valuable aid to research in many areas of human life. But it is just there, in the application of the theory of evolution to the explanation of the human situation, that a danger exists, one all the more threatening for passing almost unnoticed—it lies in wait, hidden in the mighty shadow of the word "evolution." For in the area of research on mankind, we use the reasoning tool of the "evolutionary model" as something monolithic, whereas in reality this term designates two widely differing intellectual tools forged by questing man in two widely disparate foundries.

First, there is the evolutionary concept of organic natural science: it explains the relatedness of organisms and the succession of their forms throughout geologic time as being the result of an actual connection, as the origination of many different forms from generally simpler, basic forms. Evolutionary theory also addresses the notion of simplification! In its furthest ramification, this concept of evolution includes the spontaneous generation of the simplest organisms from inanimate organic substances, as well as the transformation of matter, the metamorphosis of stars and, ultimately, of "universes."

On the other hand, historical research uses a concept of evolution that relates only to the products of human intellectual life, to all of the continuity that shows up in the record of human activity called "history." The human appears within a chain of generations, each of which hands down the fruits of its labors to the next; therefore, for each generation a new starting point is created. Inherited changes, to which human life, too, is subject, do indeed participate, but do not play the decisive role.

Common to both concepts of evolution is the basic idea of interrelated changes occurring in a complex system conceived of as a unit: the unit of the organismic type, on one hand, and the unit of large human groups—and by extension, of all of mankind—on the other. However, both systems, for which evolution is assumed, are so different that two very different intellectual structures have been erected: the organic theory of evolution and the evolutionary thought of history. They are so fundamentally different that if they are confused, the result can only be the most serious kind of obfuscation.

The historian knows the factors operating in the evolutionary relationship that he describes; even when he investigates the particular effects of the factors in detail, he knows them in principle as tradition, through oral and written versions of existing intellectual works, and as the communal life of succeeding generations. He also knows that where selection is operating in historical events, it is not immediately related to inherited characteristics but to characteristics mainfested in the area of tradition. But it is precisely that important element that the biologist does not know of in the set of relationships he conceives as evolution. The different evolutionary theories contradict each other in the very hypotheses they use to explain relationships between forms, in the adopting of factors they regard as affecting the process. The fact of mutation is known; its role in the evolution of living forms is not in dispute. But the question of to what extent the mutations we know about elucidate the transformations yet to be explained remains as open as ever. What we see, then, is that the very element that gives historical evolutionary thought its special strength—the factor that brings about the evolutionary relationship—is in biological theory uncertain and controversial. Moreover, in spite of the uncertainty with regard to the factors operating in organic evolution, we are absolutely certain that in the realm of biology these influences are very different from the forces that create the historical evolutionary context. That is one of the few, large certainties a biologist can count on in a very obscure area of research.

The validity of the concept as it is used in biology has already ceased to exist at the point where we find the facts of historical life in operation. The biological concept of development would have us believe that man emerged as the result of an organic evolution; as to the "how" of this process, the currently accepted version of the theory has nothing certain to say. And, with the assertion that the organic form "man" exists, the usefulness of the biological evolutionary concept ceases to exist. For the riddle now lies behind us (we can consider it solved or not), and we go on to speak of things that paleontology alone can never reveal: we speak of a being that has at its command verbal language and tradition. Neither the relationships among the various types of pithecanthropus and prehistoric man nor those among contemporary races have been explained by biology. This has not been due to lack of evidence, but to a fundamental difficulty: every finding of prehistory or of research into race does not lend itself to explanation through the theory of organic evolution alone, but must be understood primarily through the explanatory methods of historical research. Every fact that the biologist working in this field would like to explain by ascribing it to the descent of one human type from another and to organic advancement, the historian understands, often more accurately, to be a result of migration, trade, miscegenation, and so on.

In the field of prehistory, the two different concepts of evolution necessarily meet. And because prehistory, particularly its early phases, has been studied primarily through the methods of natural science, it has been especially easy for the biological concept of evolution to gain the upper hand. The biological concept favored the arrangement of fossil evidence in formal evolutionary series in the paleontological sense. The biological model also saw to it that cultural artifacts were arranged according to an organic evolutionary model and that such serial arrangements would be interpreted without second thoughts as the continuation of anthropogenesis. Dominance of this concept of evolution has resulted in the general acceptance of a notion of progress grounded in biology. Extrapolating the validity of this concept of organic evolution from the realm of anthropogenesis to that of historical relationships has led to every conceivable kind of overgeneralization and hasty conclusion. Curious romantic notions of the organic growth of culture, far afield of their origins, have worked to strengthen the idea that the increase in cultural objects represented a continuation of the developmental course of the human form from animal forebears. The equating of completely different concepts of evolution has reinforced the illusion far and wide that we are learned today in the most significant processes of human development.

Only careful delineation of the parameters of both concepts of evolution will make it clear to us once more in what uncertain twilight the origin of the human remains. As biology and history penetrate the darkness of prehistory, we should not have too many expectations for their meeting—we should not imagine a great moment, as when men tunneling through a mountain finally pierce the final thickness of stone that separates them and shake hands. This notion of an encounter is feasible only for the intellectual continuum of human activity. In reality, both types of research lead to a zone of darkness and silence, the extent of which no one knows.[1]

Point of View

NONE OF us investigates human existence impartially. No one forms his image by piecing together the results of his analysis against a blank background; rather, we fit our findings into a picture many of whose outlines, often scarcely visible, already exist due to past intellectual activity and the force of opinion around us. A person living in a distinctly Christian era would find a different format for his work than one living in a materialistic era; and different still from what a believer in progress, from the early days of the theory of the origin of species, would use, and from what the recent sons of the believer in progress—steeped as they are in the pessimistic theories of mental illness—would use. Scientific work resembles artistic creativity in that it starts off with a sketch, which subsequent work continually modifies. Some traits are caused to disappear and others to gain in value, but many of the basic characteristics of the sketch survive all change.

In light, scarcely discernible strokes we rough in the outline of our human, the contours given shape by thoughts about the overall distinctiveness of the human species. What we draw is the result of our looking at this peculiar "being with a history" in its entirety. Above all, we make our picture without the help of guidelines that a particular, biologically identifiable semihuman would provide: no one knows how much of the totality of the human being will be revealed through the use of the biological method. Our work is guided by the certainty that even that which can be understood through biology is essentially determined with the help of that aspect of man that must be studied with methods other than those of experimental biology.

Turning to the complete, entire being as the foundation for every kind of research—including biological—on man requires that biologists have a clear appreciation of history as being an essential fact of our existence. Against such a backdrop, even prehistoric evidence appears quite different than it does when viewed, as it so often is, in the context of organic evolution. A much better account of the extent of unknown cultural conditions, which could not have been preserved, can be given. A find of bones will not simply be considered in connection with, as it were, some wild creature; nor will a sort of paleontological sequence of human stages be constructed just on the basis of a number of stone tools. Already, there is much tangible evidence of a rich world experience for humans under very early conditions; in the future we will try much harder to represent the eternal secret of the intellectual and spiritual life of early human groups at least as something that was "available," instead of concluding summarily, based on lack of evidence, that the early human's mental state was inferior.[2]

As with many an odd idea nourished by biological arguments,

this one could never have grown if biology, in its own realm, had proceeded from a more comprehensive focal point. The negative valuation of the intellect as compared with feeling and emotion would never have captured our thoughts so thoroughly if biology had not abused its authority by using animal situations as points of departure for understanding the human condition. Only thus was it possible for new political theories to begin to place such special, high value on man's instinct, reflexes, and group behavior that merit was ascribed to the "healthy predator"; to adopt numerical growth of a group as proof of its value; and to establish animalian innocence as the kind of clear conscience desirable for a human society.

All these positions deriving from biology probably owe their significance in part to the perplexity of leaders, but also to the mounting appreciation, furthered by the theory of evolution, of the animalian state as the norm by which the human "condition" is to be understood. However, evaluated according to such a norm the human lacks the sureness of instinctive judgment; it lacks the unexceptionable matter-of-fact behavior that goes with preservation of the species. And the human being lacks the much vaunted brutish innocence in killing and imposing his will.

Thus, it is a grave error to believe that the basis for evaluating human existence can be found with certainty by studying animal behavior. The biologist will indeed provide the basis for important comparisons, by means of which our singularity will stand out more clearly, but understanding of these biological particulars will be only part of our effort to find, in the facts of our own human existence, the laws according to which we lead our lives.

The studies presented in this review are but fragments. They refer only to separate details, by means of which I hope to give fruitful points of departure for a new conception of the biologically accessible aspects of human existence. But there is another reason for their being called fragments: they have to do with matters that are comprehensible biologically and yet have been compiled in the certainty that only a view of the more comprehensive reality will lay the groundwork for a new image of man—and image that, in the future, will have the power to shape the human endeavor.

The material in the following pages is presented in hopes of making a contribution to the future image of mankind.

1
The Newborn Human

IT IS open to discussion whether biological research or another kind is the method for probing the nature of the human—but the science of the living has been our own method for problems of anthropology. Constantly renewed attempts using comparison to place the developmental modes of the various groups of vertebrates within a larger context have eventually led to new research on human development. These remarks will also use comparison as their point of departure, comparison of those juvenile stages of higher mammals with those of humans.

Altricial and Precocial Young

THE HELPLESS newborn human reminds us of similar developmental states in mammals and birds; an inner bond of understanding makes the animal mother seem more human to us, more closely related than the animal would otherwise appear to be. This impression of accord goes so deep that it is scarcely noticed how unusual the nature of the human baby actually is, how much it deviates from what is the rule for higher mammals. The characteristics all mammals have in common—loving care of the young and nursing—have diverted our attention from the differences in the ways the various groups develop. To address this matter, we must first look clearly at the special features of human newborns.

The extremes of state at birth that we see in blind kittens and frisky colts are not distributed haphazardly among the individual categories of mammals; instead, within any mammalian group there is a constant relationship in evidence between the degree of organization and the way a mammalian group develops. The development of mammals whose body structures show little specialization and whose brains are only slightly developed is usually characterized by short periods of gestation, a large number of young in each litter, and the helpless condition of the babies at the moment of birth. In these early stages they are usually hairless, their sensory organs still closed, and the temperature of their bodies still completely dependent on warmth from an external source (insectivores, many rodents, and small carnivores—the marten, in particular). In birds, too, we know of such naked early stages with closed eyes—song birds and woodpeckers, for example, and the term customarily applied to birds, *Nesthocker* (nidi-

---

Figure 1.1a. Marsupials. Of all the mammals, marsupials are the most fetal at birth. A newborn opossum *(Didelphis),* after a fetal period of twelve and a half days, looks just like the infant giant kangaroo in the illustration after a thirty-eight-day period of development. Eyes and ears are sealed and not visible from the outside, the mouth slit has grown together at both ends, leaving only a round mouth opening with which to attach to the mother's teat. The tongue is already keel-shaped, formed for an extended suckling period. The forelimbs are much further developed than the hind limbs and bear claws. Their task is to enable the newborn to get from the birth canal to the pouch, a feat it must accomplish alone. Photographs: Dr. F. Müller, Zoological Institute, Basel.

Figure 1.1b. Golden hamster. Some rodents, especially the golden hamster *(Mesocricetus auratus)* are at birth quite similar to marsupial offspring. They are born in a nest and, unlike the marsupials, do not have to exert themselves at all. Correspondingly, both their fore- and hind limbs are evenly developed but still quite embryonic at thirteen and a half fetal days, the number of days at which an opossum is born. Hamsters are born on the sixteenth or seventeenth day. The picture on the left shows the still incompletely sealed eyelids and ear openings: the one on the right shows the moment of birth, when the eyelids are completely sealed and the auricle is folded forward and overgrown with skin. Development of the coat is delayed. The light spots are hair follicles still embedded beneath the surface of the skin; at birth, only the follicles of a few special sensory hairs on the head appear, as bumps. Photographs: Dr. F. Müller, Zoological Institute, Basel.

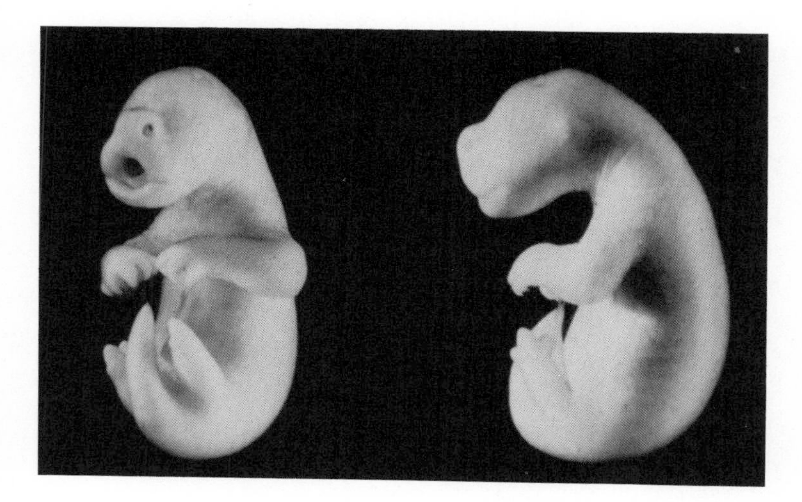

Figure 1.1a

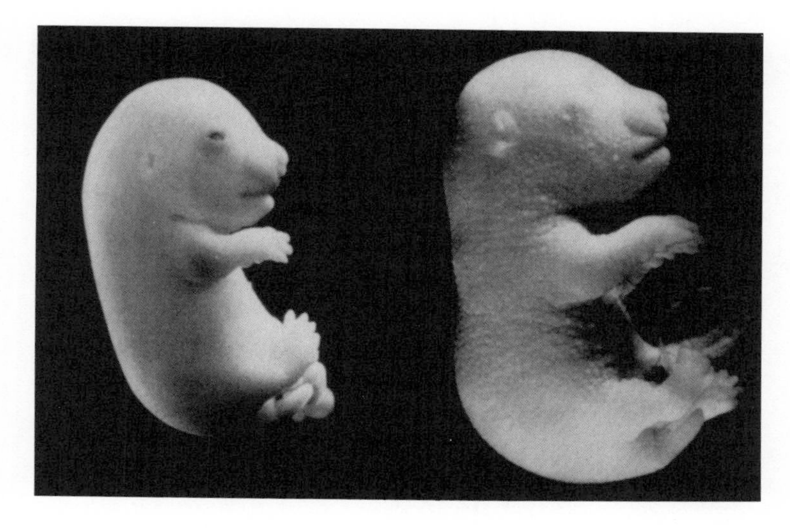

Figure 1.1b

colous, altricial),* has come to be applied to all similar developmental states, even in mammals. A completely different type of development is shown by more highly organized mammals, whose body structures are more specialized and whose brains are more complex (ungulates, seals, whales, prosimians, and apes). For these creatures, development within the uterus lasts quite a while, the number of young in each litter is reduced to two or one, and the newborn are well developed, appearing much like adult animals in both form and behavior. Again, chickens, ducks, snipe, and other similar birds produce well-developed young, and the usual term applied to the latter, *Nestflüchter* (nidifugous, precocial),* is also extended to cover the corresponding developmental state of mammals.

*Throughout, I have used the terms *altricial* and *precocial* respectively for *Nesthocker* and *Nestflüchter* (literally "nest-squatter" and "nest-fleer").——Trans.

TABLE 1.1. Ontogenetic Relationships in Mammals

	Low Level of Organization	*Higher Level of Organization* †
Gestation period	very short (e.g., 20–30 days)	long (more than 50 days)
Number of young per liter	large (e.g., 5–22)	usually 1–2 (rarely 4)
State of young at birth	altricial	precocial
Examples	many insectivores, rodents, and marten-like carnivores	ungulates, seals and whales, apes and prosimians

† Many mammals with extremely specialized organization but with a relatively low level of brain development behave like the more highly organized mammals, for example, sloths, anteaters, and bats. Our table only calls attention to relationships that are at times too little noticed and which are currently the object of evolutionary research.

What a contrast there is between the torpid suckling stage of a squirrel or European polecat, whose eyes do not open until about thirty days after birth, and the quick alertness of the newborn elephant, calf, or giraffe, each of which can stand and even follow the herd just hours after birth! Cats and dogs fall somewhere between these two extremes, a position that also corresponds to

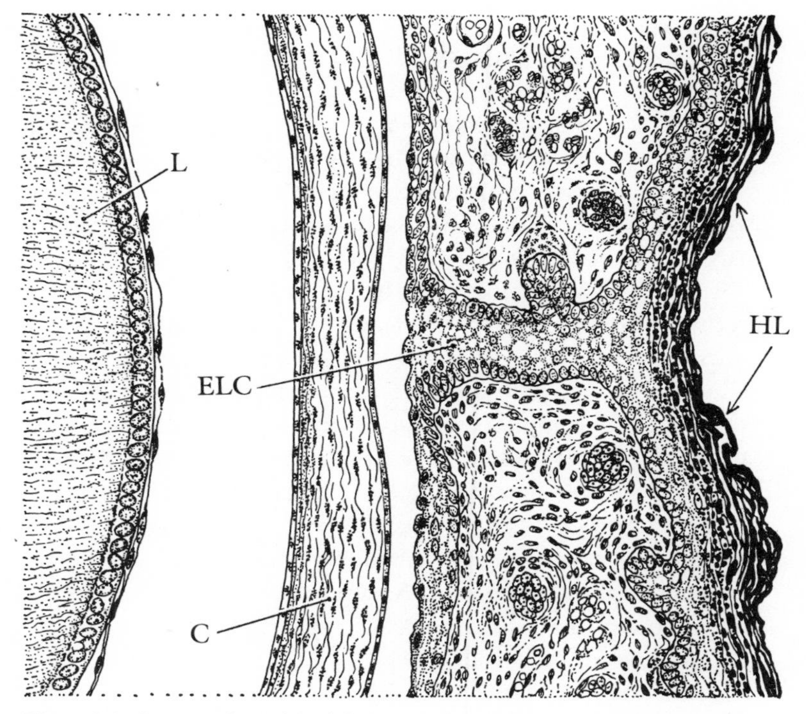

Figure 1.2. Cross section of the lid closure of a newborn rat. Both lids are completely grown together for a time, whereas in altricial birds they always remain separated by a horny layer and are sealed only superficially. The closure of the lids in marsupials is accomplished by means of a somewhat different kind of tissue structure from the growth found in insectivores, rats, and mice. ELC = epithelial lid closure; C = cornea; HL = horny layer of the skin (stratum corneum); L = front of the lens. Figures 1.1 and 1.3 show the stages of lid closure.

their degree of specialization and level of organization: the young are indeed altricial but at the moment of birth are already much more developed than baby rats or hedgehogs. It is no accident that with dogs and cats the number of young also falls between the two previously mentioned extremes. (This applies only to wild forms, for the number of young produced by domestic animals can be increased as a result of selective breeding, as we know from dogs and swine.) Table 1.1 summarizes the most important ontogenetic correlations.

Thus far, we have only been able to verify that such correlations exist. We are striving to clarify these relationships, which in many instances are so inadequately understood, because they contain information about the human situation. Our attention is first drawn to the second group of correlations in the table, that of the higher mammals, which includes the related group of the highest mammals, the so-called primates, and, with them, the human.

It has been little more than half a century that we have had precise information as to the length of gestation times in primates.[3] A chimpanzee was born for the first time in captivity in 1915, and it was only during the most recent decades that the number of observations has increased to the point that we can now say that the average gestation period in chimpanzees is 253 days, a length of time corresponding to that for humans—280 days. If we say that in man the actual period of time from fertilization to birth averages 266 days, the same statistic for chimpanzees is 237 days. A gestation period of about 275 days has been established for the orangutan and, according to current reports, one of between 251 and 289 days for the gorilla.

It was not until 1956, in Columbus, Ohio, that a gorilla was born in a zoo. Two years later, Goma, the female gorilla in Basel, became famous, and since then our experience with these animals has steadily increased. Furthermore, in recent decades, observations in the wild have made important advances.

We have long known that all primates have a small number of young. From the lowest level of prosimian organization to the great apes, we find "human" situations: one baby, rarely two; only in the marmosets of South America is the birth of twins common.

Newborn primates are precocial. They are all born with open eyes and well-developed sensory organs and, beginning at the earliest stage of life, are capable of all kinds of movements. The babies of a few species of prosimians do not open their eyes until just after birth, but, in general, the early independence of small prosimians—lemurs—is striking. But the offspring of guenons, macaques, and baboons are also so independent that no one who has observed these creatures at the time of their birth would hesitate to label them precocial.* As soon as they are born, they experience a special compulsion to cling, gripping their mothers' furry hides with their strong hands and feet. Even though the mother sometimes lays a protective arm around her baby or holds on to it now and then with her hand, the baby supports itself, and does not even lose its grip when the adult goes swinging through the branches in bold leaps. In calmer moments the infant clambers around a bit on the mother; she is the first "tree" of its young ape's life, and for a long while yet, she will pull the youngster sharply back by the tail when it tries to get too far from her. The clinging instinct of young apes is an appropriate defense mechanism against the dangers of life in the trees; this instinct completely dominates the baby's early behavior, concealing the great capacity for independent movement that is actually present in the newborn primate.

When we realize what a strong compulsion this drive exerts, requiring as it does that the hands and feet be so completely in the service of holding on, the freedom a newborn human has with its hands and feet seems quite a contrast—all the more so since our baby is so much more helpless than the baby ape. The free play of the limbs, which gives our infant possibilities so much richer than those available to the newborn ape, reminds us that our birth condition is not merely "helpless" but also entails important freedoms.

* Portmann uses *Nestflüchter,* and the next sentence begins, "The term is, of course, somewhat misleading, for they do not actually flee a "nest" but rather . . . ." I have omitted these words.——Trans.

Comparison of our birth state with that of other mammals must be based on the infant forms of New World primates and of guenons, baboons, and related species. At first glance, this might seem strange, for it appears clear that the initial stages of juvenile development in anthropoid apes are more closely related to our own. Briefly, then, the choice of our starting point must be justified.

A few decades ago, the view was still widely held among zoologists that the decisive phylogenetic separation of the human type was accomplished fairly late, toward the end of the Tertiary, and that consequently, phyletic forms like our present-day anthropoid apes impart a good idea of the ancestral situation. However, as early as 1925, widely divergent opinions were voiced, postulating a very early evolutionary separation of the hominid type, that is, the separation of the human line from that of the Pongidae, whose last representatives are the great apes we are familiar with. At the time, these voices went almost unheard, but research during the last four decades has clearly established this view as the dominant one. It must have been about twenty to twenty-five million years ago, toward the end of the period geologists call the Oligocene, that within groups of monkeys related to the guenons and macaques the paths of the two great evolutionary lines separated. We

---

Figure 1.3. The development of the shrew. Insectivores are a group of geologically very old mammalian forms, which, like the marsupials, have retained most of the early developmental characters of the entire class of animals. The photograph, taken on the twenty-first developmental day of the house shrew *(Crocidura russula)*, shows the early state with eyes open, the lids not yet formed. The auricle still stands out from the body at right angles. In the fetus—almost ready to be born—of the forest shrew *(Sorex araneus)* the lids have grown over the eyes, sealing them shut, and the auricle is already folded more definitely forward. In the first picture, the five rays of the fore and hind digits are still embryonic; the second picture shows more differentiation. The sharp bulge of the forebrain is conspicuous in the early fetus; the mouth slit of the late stage is much closer to what is typical for the species. The picture of the just-born forest shrew (below) shows how embryonic and helpless this altricial infant is when it is born after a gestation period of twenty days. Photographs: P. Vogel, (graduate student) Zoological Institute, Basel.

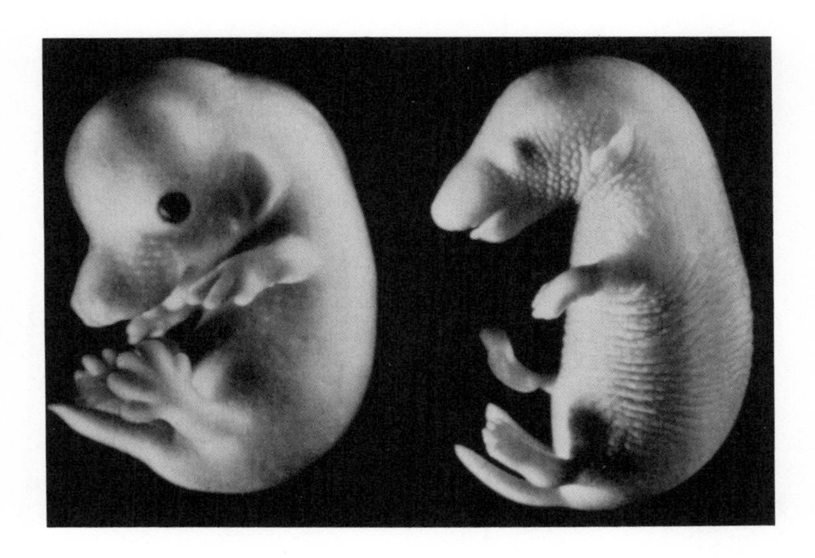

Figure 1.3

have little concrete information about the primates of that time, but scientists who study those extinct forms are today in widespread agreement that the evolution of the human life-form began in geologic time with a still unknown, but probably very form-rich transformation that was relatively independent of the evolutionary path of the great apes. The decisive event must have been accomplished over a period of ten to fifteen million years, a *Hominisationsfelde,* or period of "hominization," as this spatiotemporal continuum of our emergence is sometimes called.

These new insights are of great significance for our concept of the great apes. They are probably very closely related to humans but not true ancestors, with the consequence that all forms that might seem to be prototypes of our form of life are not *prehominids* but members of another line with related traits—*parahominids.* This does not reduce the value of the evidence provided by the study of the great apes, for they are, of course, the only phenomena that, in a certain sense, furnish us with models for understanding, models for the human prototype, which we will never see alive. Knowledge of contemporary great apes, applied with great circumspection, can still supply the image of those prototypes with some features.

However, the prerequisite for such a comparative evaluation is exact knowledge of the behavior of the species still sharing the earth with us, and it is precisely the largest of these whose existence is now threatened. Whereas the already mentioned early evolutionary split between the great apes (Pongidae) and humans (Hominidae) was recognized as early as 1925 and was particularly clear after 1945, it has been only in the last decade, really since 1960, that the infancy and maturation of the different groups of apes have been thoroughly understood. It is important to take a look at these circumstances and to assess which of them are essential to the explanation of the emergence of our own particular natures.

Just as firm as the concept of the gradual evolution of primates and humans is the assumption that the most distant ancestral forms were altricial in their infancy, in the manner of insectivores. Whether the development of the east Asian and Indo-Malaysian tree shrews (Tupaiidae) offers a picture of a first step along the path toward the primate level is in dispute. Tree shrews exhibit an archaic mammal form with altricial offspring; the number of babies in each litter has already been reduced to between three and one, and a gestation period of between forty-three and forty-five days may correspond closely to a somewhat more complex brain structure. At the outset, these baby tupayas are helpless; their eyes do not open until they are fifteen days old. Subsequent development is swift; after five months, the size typical for their species is reached, and one month later, tree shrews are sexually mature. This course of development fits in quite well with the overall picture that biologists must make of the precursors of primates; but, as already mentioned, observations of this sort can only serve as an extremely general model of the development of that unknown mammalian group.

Prosimians have taken the step beyond the old altricial level. After just five months, the lemur begins life on its own; we know that the mouse lemur is capable of reproduction when it is seven to eight months old. At the ape level of primate evolution the period of time during which the infant is closely tied to the mother lengthens to more than one and one-half years in South American howler monkeys, and to one year in macaques and Old World baboons. A longer period of sexual immaturity creates an extended juvenile period, which determines life within the group often on into the fourth year, even into the fifth year for many male macaques. Still more years may pass before the offspring achieves full status within the group (table 1.2).

The formation of the great ape and hominid types entails a further lengthening of the juvenile period: three years of the closest bonding between mother and offspring. Then, in the gorilla, there is a two-and-one-half to three-year period of extended adolescence and, in the chimpanzee, one of seven and one-half to eight years. For chimpanzees and male gorillas, complete social maturity is delayed until on into the tenth year.

It is not surprising that, with such a prolonged adolescence,

TABLE 1.2.

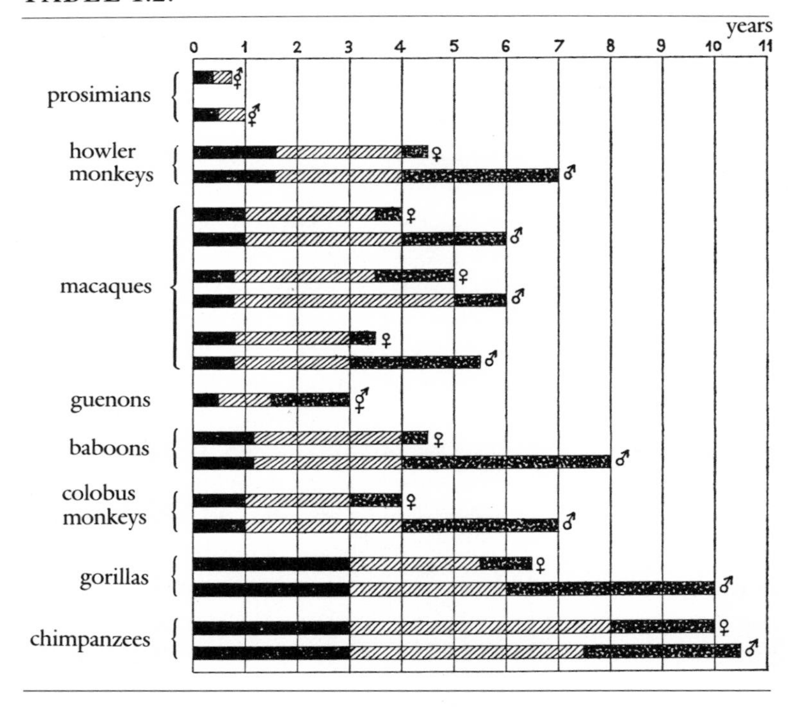

NOTE: The data in this table came from studies that were collected by De Vore (1965; see note 3). From top to bottom: prosimians (microcebus and lemur), after Petter; howler monkeys *(Alouatta),* after Carpenter; macaques: *Macaca rhesus,* after Southwick; *M. fuscata,* after Imanishi; *M. radiata,* after Simards; guenons *(Cercopithecus ascanius),* after Haddow; baboon (several species), after Hall and De Vore; colobus monkeys or langurs *(Presbytis entellus),* after Jay; gorilla *(Gorilla gorilla beringei),* after Schaller; chimpanzee *(Pan satyrus).* The three age levels: ▬ childhood; ▨ juvenile: ▩ subadult (from onset of puberty).

dependence of the offspring on the mother increases. The great apes exhibit differing degrees of this needfulness immediately after birth. Nevertheless, the basic traits of the early independence of the primate baby are unmistakable: early ability to grasp the mother's fur and cling to her by itself is more or less evident. We should not be too quick to assume that behavior exhibited in zoos is typical, particularly in the case of the offspring of great apes raised by humans. G. Schaller has also warned against such assumptions after having spent months observing gorillas in the wild. Thus, in our review of parahominid traits, we shall consider carefully the development not just of gorillas and chimpanzees but of all anthropoids, ever mindful of the fact, however, that the unknown, solitary human path compels us to leave many blanks in the outline of our own phylogeny.

Similarities between the offspring of the great apes and our own are so evident, particularly in the early stage, that I am always astonished at how much keen perception is expended in documenting these details of congruence. This is as unnecessary as is the assertion that we really are related to the apes. This relationship, which no one doubts, seems to make sense but in its particulars is full of puzzles. Thus, the similarity between youthful forms is still no answer to the question of our origin, and so, in our analysis, we shall try to single out and examine closely not just similarities but contrasts large and small. Perhaps this way of proceeding will show us more about man's origin than the continual emphasis on parallels.

Morphological Features of Our State at Birth

EVERY COMPARATIVE classification of newborn humans must proceed from the striking fact that the group of mammals from which both humans and the great apes have emerged exhibits a secondary birth state for mammals. The form of the newborn is like that of the adult in bodily proportions, particularly in the limbs, and the infant is born with the eyes open.

There is a little noticed but significant sign that, at birth, the developmental stage of primates must be termed precocial: the fetal development of the sensory organs! To be more precise, within the womb, all primates go through a stage during which the outer sensory openings are closed.

Why the eyelids, the ear passages, and the nasal openings close during the fetal stage in the womb and open again just before birth cannot be explained functionally, for during this period of time there is no change in the living conditions in the womb that would affect these epidermal structures. In contrast, such closures make sense for altricial infants and can be regarded as preparation for an early birth; the closures provide the still undeveloped sensory organ with the necessary liquid environment and protect it from exposure to air. Thus, closure provides for the peripheral part of the sensory organ what a special sac, the amnion, provided at earlier developmental stages for the whole germ: the protection offered by the primary, aqueous environment.

It is not just the eyelids that close protectively over the as yet unfinished visual organs; a similar closure shields the middle ear, where one of the most curious transformations known to vertebrate research takes place: the metamorphosis of what were the articulating elements of the jaw of the reptilian ancestor into the sound-conducting apparatus of the auditory ossicles.

The closure of the nasal openings during fetal life is of a completely different order than that provided by the protective membranes covering the eyes and ears. Nasal closure also appears in reptiles, during whose embryonic stage the eyes and ears are never closed to the outside. The reason for this is still completely unknown.

As has already been said, the formation of such sealing structures a considerable length of time before birth cannot be explained functionally. It seems to be an event rooted in the genetic material of mammals. In their early stages, the higher types follow the developmental plan of the group, going through the same steps as the simpler forms.

Since this developmental step of the closing and opening of the sensory orifices while still in the womb has no other significance, it can undergo considerable alteration or mutation without producing an adverse effect. As a matter of fact, we find many small variations, be they in the form of the closure or in the time the opening takes place, which can probably be attributed to mutations that did no harm to the germ.

Primates still in the womb go through a stage that, with regard to form, is comparable to the birth state of an altricial infant. Therefore, by the time a primate is born, its growth has arrived at a more advanced developmental stage.

Humans, too, while still in the womb, undergo all these changes in the sensory organs, changes that are characteristic of altricial infants. The most noticed change is the sealing of the eyelids, which grow together during the third fetal month and open again

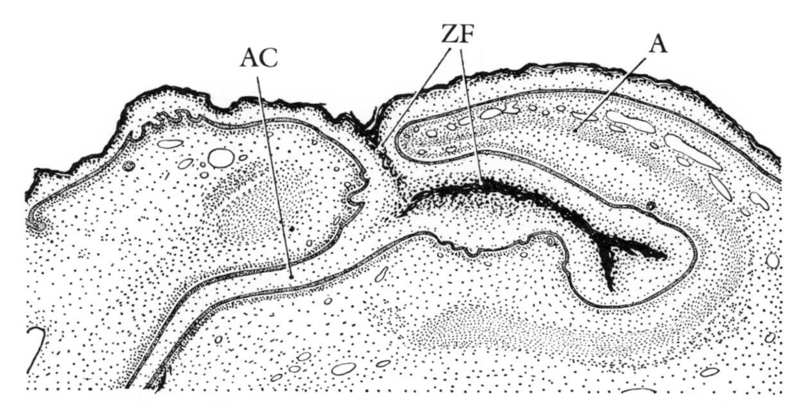

Figure 1.4. Horizontal section through the ear region of a newborn rat. The auricle is folded down toward the front (toward the left) and completely fused with the skin of the body. The outer auditory canal with the ear drum area (not visible here) is protected. AC = auditory canal: A = auricle: ZF = zone of fusion.

at the end of the fifth month. If humans were born as simple altricial infants, something like the helpless young of a squirrel or a marten, they would have to be born at about this five-month developmental stage. In reality, however, humans mature further in the womb to the precocial stage of colts or calves, with open sensory organs and well-developed locomotor systems; they attain a level of formation that is characteristics of all higher mammals.

The insight that the helpless newborn human is actually, quite inconspicuously, a kind of precocial infant (although definitely atypical) puts in their proper place many facts about our makeup that must have appeared formerly as curiosities in the bodily structure of an altricial infant. For instance, it has been emphasized that in the newborn human, the structure of the medullary sheath of the corticospinal tract of the central nervous system is much more like that of a foal than that of any truly altricial infant, as a cursory glance will prove. In humans, the medullary sheath begins to form in the ninth fetal month and, as in young ungulates, is quite far advanced at the moment of birth, whereas with true altricial infants, young rats or mice, for example, development of this structure does not begin until some time after birth. In the same connection, it should be mentioned that the number of motor fibers in the ventral roots of the spinal cord, in the region of the

---

Figure 1.5. Embryo of a guenon. (After Buffon's *Histoire naturelle,* 1789.) Throughout the ages, there has been an active certainty of our puzzling relationship with all forms of apes. Count Buffon had the fetus of a guenon drawn for his *Histoire Naturelle* (the 1789 edition did not appear until after his death) to call attention to this relationship. In the picture, he captured the significant moment in the course of development when the higher mammals—including the human—while still in the womb, pass through the state of altricial infants at their early birth: the phase in which the eyelids are closed at the same time that the brain is prominent, a period during which the similarity to our own early form is particularly strong. The sensitive choice of position and the arrangement of the drapery reflect the spirit of the time, when the ape was held to be a being very close to us; the anthropoid ape was even called *"Homme des bois,"* our curious brother from the forests. It was not without reason that Linné called it *Homo nocturnus,* or man of the night, to contrast it with us humans, the "men of the day."

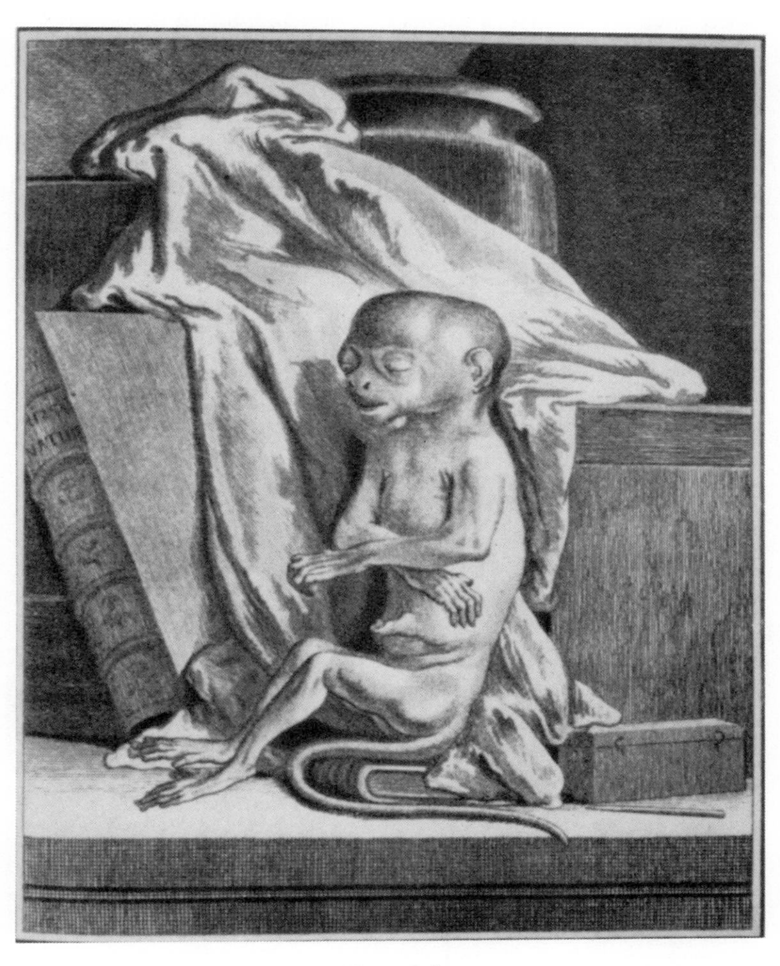

Figure 1.5

arms as well as in that of the legs, is the same in the newborn as in the adult. In early developmental stages, the area toward the head is clearly further along in formation than the posterior part of the body. This is why in true altricial infants, which are born earlier, the formation of nerves has advanced further in the forelimbs than in the hind limbs. The human, however, behaves like a precocial infant, both of whose limbs have attained the same degree of development. It would be desirable for observers to be more open to the subtler signs that testify to the mental developmental level of the newborn human. Our insight will continue to be sharply restricted if we persist in the opinion that we are only dealing with a meagerly developed altricial infant. The varied, careful studies of Stirnimann refute this point of view and pave the way for some essential explanations.[4]

The morphologically demonstrated affiliation of newborn humans, and of all infant primates, with the level of the precocial type is further emphasized by the fact that the composition of human milk is much more like that of mammals with precocial infants than, say, dog or rat milk is.

Figure 1.6. Comparison of the development of a primordial mammal having early birth as an altricial infant *(below)* with the development of the higher mammals with longer gestation periods and closure of the sensory organs during the early period *(above)*. The black background indicates the time within the mother's uterus.

The Proportions of the Newborn

BEYOND THE first sensory impression, the powers of observation cannot penetrate the hidden nature of things without having recourse to preconceived ideas. For that reason, it would often be of more value in scientific accounts if the scientist, instead of claiming objectivity, would clearly indicate the "prejudice" that contributed to the advance into unknown territory. We assert once again that, in the following, the newborn human will be considered from the morphological standpoint as a creature affiliated with the precocial mammals. This can be only one of many perspectives of the prob-

Figure 1.7. Drawing of a gorilla fetus near the moment of birth. Photographs of newborn gorillas do not always show the proportions clearly due to the pelage, but this drawing clearly shows the length of the arms compared to that of the legs, the mark of adult proportions.

lem under consideration, but it will help us to see the most striking trait of our birth state—helplessness—in its true context: not as the primitive, somatic immaturity of an altricial infant, but as a very exceptional situation within the mammalian group. In fact, with respect to the full precocial type, humans are secondarily altricial.

Just the study of bodily proportions clearly brings out the uniqueness of our babies. From birth on, the young of higher mammals maintain bodily proportions close to those of the adult forms. Thus, on the first day of their lives, foals, fawns, young whales, and small harbor seals are already miniature versions of their parents, with the only particularly conspicuous difference being the relative size of the head (and the brain); in the same way, the newborn ape is similar to the adult in the size ratio of limb to torso. Anthropoids also follow this rule. The long limbs of orangutan or gorilla fetuses are immediately conspicuous. If fetuses of different ages are depicted as being equal in overall length, such a series exhibits exactly the same length ratios between torso, arms, and legs even at very different ages; the only difference is that, as expected, the head's share of the length steadily decreases with age (Babor and Frankenberger; see note 5), for in early stages, the brain is particularly far ahead in development.

How differently the newborn human is built! How different are our limbs from their ultimate state! Comparisons based on growth provide a useful, numerical expression of this important difference between anthropoids and humans. We take the length of a part of

TABLE 1.3. Comparison of Lengths (at birth : at maturity)

	Chimpanzee	*Human*
torso	1 : 1.95	1 : 2.65
arm	1 : 1.69	1 : 3.29
leg	1 : 1.69	1 : 3.94

the body at birth as the unit with which to compare the same part of the mature body. If the ultimate state is a proportional enlargement of the form at birth, then the growth values of the part remain relatively close. This is true for the torso, arms, and legs of chimpanzees (table 1.3), whereas in humans, the corresponding values are much further apart.

Human torsos, arms, and legs must each cover a completely different developmental trajectory to reach the size of the adult state; the proportions in the human newborn are completely different from those of the mature form. This much we can tell just by looking. Such differences, however, have attracted too little attention because, even in baby apes, we primarily wanted to see what could be interpreted as a preliminary stage of anthropogenesis. Schultz[5] measured in detail the changes in body proportions in apes and humans. He compiled the relative lengths (indices) for eleven different body measurements of a fetal stage that, with regard to form, corresponded to the fifteenth week in human pregancy. He compared these fetal values with the corresponding values for the mature forms and determined the difference between an organ's two indices. When the difference thus arrived at is expressed as a percentage of the fetal index, the result is a value that expresses the deviation of the fully developed organ from its early state. If animal forms grow in a similar way, these figures would have to be close together. From all of the eleven measurements compared, Schultz calculated a mean deviation and came up with the following figures: orangutan 6.7; gibbon 7.7; chimpanzee 7.9; gorilla 9.8; and for the human 23.4.

Of course, this has nothing to do with the state at birth, but because the early fetal stages resemble each other much more than the mature forms do, the large, one might even say surprising, difference that so clearly sets humans apart in this series is primarily due to late changes that set in after birth. Schultz's figures provide a vivid indication of the special development of humans after birth: apes arrive more quickly at adult proportions while still in the fetal state, thereby resembling the higher precocial mammal type, in which the birth state is a miniature version of the adult form. However, in humans, genetic factors—the particulars of which we do not yet understand—keep us from such early attainment of the bodily proportions typical of our species; humans attain mature proportions late, after birth, and only after peculiar transitional phases in a mode of growth different from that of all apes. I hope further research will show that the light in which Schultz's results appear here will illuminate an important aspect of the human being and its development.

Schultz has already indicated that his findings contradict a theory that was long and widely respected: the fetalization concept of Bolk,[6] which seeks to prove that essential characteristics of the human form are based on the retention of fetal proportions. Again and again the point is made that the juvenile stage of anthropoids bears traits that are particularly human. This view is a consequence of looking exclusively at the head, where the significant size of the brain is just as striking as the suggestive humanness of the face, where the jaw still protrudes only a little. But if you look at other bodily proportions, in addition to similarities striking differences appear, in particular the deviating bodily proportions. Granted, it will probably require yet more time to accomplish the conceptual transformation and habituation needed for the similarity between the form of the head in childhood in both humans and apes to be seen above all else as a general primate characteristic—the similarity will not reflect primarily the humanness of apes, but rather our likeness to primates. Only then will the peculiarly human quality of the special bodily proportions of the newborn child become more obvious.

Sometimes, to see the human aspect of the entire body more clearly, we must shut out the forceful attraction exerted by the head. Theories of descent have often been taken up with the head exclusively: partly because we are partial to that focal point of the human phenomenon; partly because our attitudes are compelled by evidence that consists mostly of skull fragments. Perhaps the information provided by our research will encourage increased observation of the growth phenomena of the entire body. Within their area, biologists must do for a while what Rodin did when he wanted to represent the power in a stride, the intense, suppressed vitality in every bodily part, the self-declaration of the flesh—he had to disregard the form and function of the head.

Birth Weight and Brain Formation

THE CONSIDERATION of weight ratios leads us to another essential peculiarity of our birth state. The newborn's helplessness would lead us to expect that humans, when compared with the great apes, whose infants are so much more mature at birth, would also be behind in the development of mass, especially if the comparison is made with gorillas, whose body mass so far surpasses our own. This expectation is reinforced by the impression offered at first glance by postfetal development that we humans grow slowly, much more slowly than other mammals do. This would be an excellent view if at birth, our babies weighed much less than those of the great apes.

The truth of the matter is just the opposite! Newly born human infants are quite a bit heavier than those of the great apes. The figures we have for apes still do not permit an infallible comparison of averages. Nevertheless, they justify the assumption made by Brandes (1931) that the average weight of all great ape infants is about 1,500–1,900 grams. The high weights given for the orangutan (2,500 g) and the chimpanzee (2,300 g) are the highest figures available, but they would best be compared provisionally not with the highest figures available for humans but rather with average high weights after the corresponding gestation period, for it can hardly be expected that the few figures we have for great apes would include extremes. The figures in table 1.4 reflect this.

The picture is unmistakable. Whether we use the earlier-born chimpanzee for comparison, or the orangutan and the gorilla, whose gestation period is more like that of humans; whether we use average values or high weights: the figure for human babies is far above that for the great apes. Must we simply accept this unexpectedly high weight for our newborns as an irreducible fact, or does it make some sense when understood within a larger context?

The figures compiled in table 1.5 show that there is no set relationship between the mother's weight and the weight of the

TABLE 1.4.

Length of Development		*Human*	*Chimpanzee*	*Orangutan*	*Gorilla*
253 days	average	2,500 g	1,890 g		
	high	3,500 g	2,300 g		
275 days	average	3,200 g		1,500 g	1,800 g
	high	4,500 g		2,500 g	

TABLE 1.5.

	FETAL TIME*	AT BIRTH		ADULT FEMALE	
	Days	*Total Wgt. g*	*Brain Wgt. g*	*Brain Wgt. g*	*Body Wgt. kg*
gorilla	270	1,800		430	100
chimpanzee	253	1,890	ca. 180–200	400	40–75
orangutan	275	1,500		400	75
human	280	3,200	ca. 370	1,450	50–75

*Fetal time is given the way it usually is for humans; the calculated "age from conception" is correspondingly less. The weights are averages, which should show clearly the similarity among the great apes as to birth weight, brain weight at birth, and weight of the adult brain as compared with the unique situation in humans.

baby that would be valid for the whole primate group. The very different sizes of the three adult apes, from about 40–75 kilograms in the chimpanzees to the more than 100 kilograms in the gorillas, stand in contrast to their relatively similar birth weights. On the other hand, the body weight of human females is similar to that in orangutans and the lighter forms of chimpanzees, and even approaches the lower limits of the scale under consideration here if we compare the average of Japanese females, which is 49 kilograms, and that of many women of the Sunda Islands, who are even more slightly built. In spite of this similarity of mature weights, the newborns of even delicately built human types weigh about 3,000 grams, and those of other types even more—in any event, almost double the weight of anthropoid ape babies. Comparison of the birth weight with the body size of the mother provides no meaningful explanation for the conspicuously high weight of our babies.[7]

On the contrary, we get a first glimpse into the hidden relationship if we compare the development of mass of the whole body with that of one organ, which, through the complexity of its structure, influences to an extraordinary degree an animal's entire development: the brain. The more diverse the lines of communication between the nerve elements and the greater the number of cells, the longer it will take for the nervous system to reach its mature form and the more imperative it is to establish early the correct relationship between the growth of the brain and the formation of the whole body.

We shall first try to give a picture of typical mammalian brain development as it takes place after birth. In so doing, we take into consideration for these first comparisons only the ratios of mass, for if cell size is relatively similar, these figures give a useful, even if approximate, picture of the differences that exist between the birth state and the mature form. Table 1.6 compares brain weight at birth with that of adult animals of both altricial and precocial type, thus providing the "multiplier factor" for the brain mass.

True altricial types such as rats and rabbits show the highest values. If the relatively simple brain structure in these species is taken into consideration, the multiplier factor gives an all the more vivid idea of how far from its developmental goal such a brain is at the moment of birth and how immature this developmental stage really is. The multiplier factor for cats indicates an organization that at birth is somewhat closer to the final form, which corresponds to the already emphasized intermediate position of this terrestrial carnivore. The figures for ungulates and primates clearly show that, in precocial types, the brain at birth is quite close to its mature state.[8]

Knowledge of the multiplier factor of a mammal's brain after birth has become an important aid in evaluating the developmental type and has also enabled many phylogenetic assertions to be made: assertions that, if research were limited to fossil finds, would either not have been determined at all or would have been very uncertain. The values given in table 1.6 were gradually established during the years 1950–60 and must, of course, be supplemented. Still missing are figures for seals, toothed and toothless whales, and elephants. We can already venture some predictions today, but that is not our task here.

One of the first results is the limit represented by the multiplier factor 5: everything above that belongs to the primary altricial

---

NOTE: Brain growth is a particularly important aspect of individual development. The multiplier factor tells how much the anterior brain increases in mass after birth; it conveys an idea of the developmental state of this important organ at the beginning of postfetal life. The table summarizes the most important of these factors obtained from studies done by K. Mangold-Wirz. The solid lines separate large related groups. Crosshatched rectangles refer to the situations most likely for the immediate ancestors; the black rectangles give the values for today's mammals (the exact figure is at the right). The arrows indicate the direction of evolution: development from altriciality to precociality (= arrow pointing right); evolution in the opposite direction (man), from precociality to secondary altriciality (= arrow pointing left) and, for isolated carnivores, to particularly extreme altriciality (bears). In the list of names, the species with altricial young are on the left and those with precocial young are on the right: the multiplier factor 5 is the limiting value.

TABLE 1.6. Postpartum Multiplier Factors of the Cerebrum in Mammals

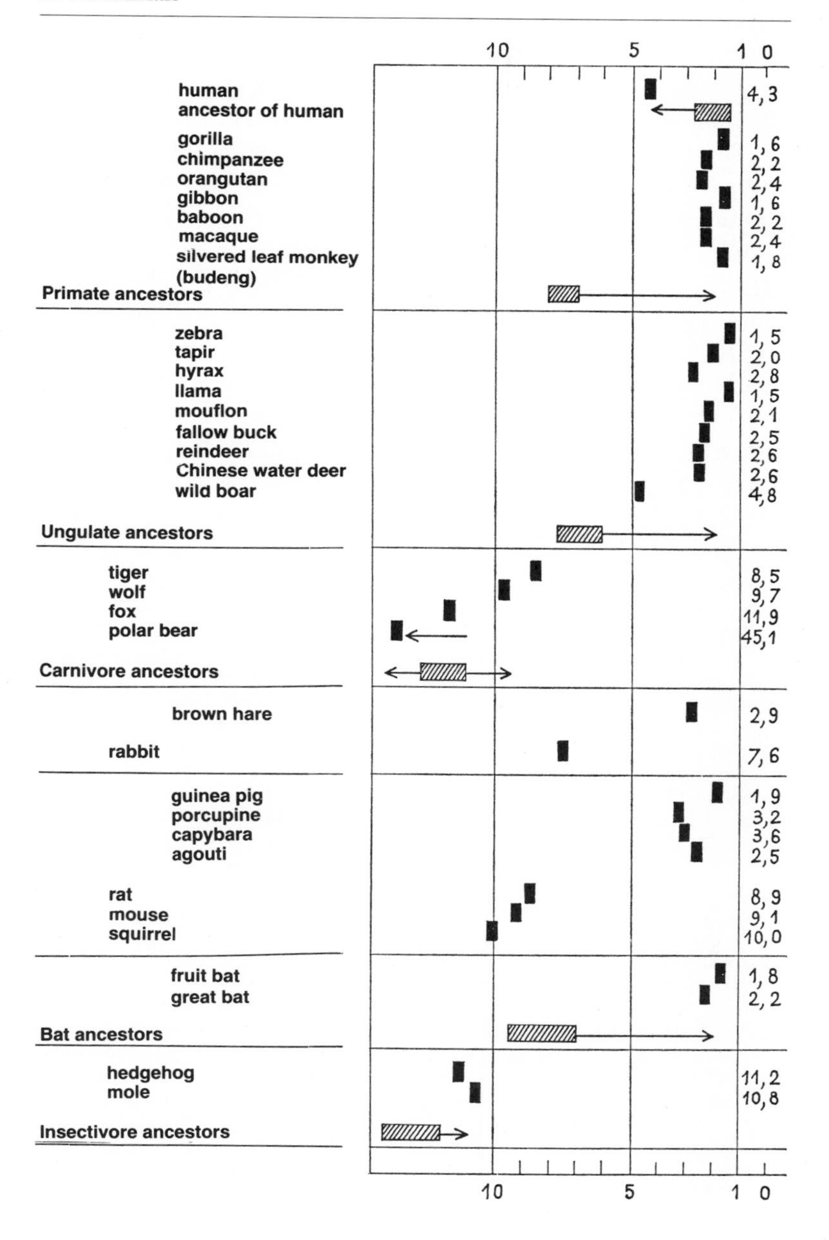

type, and everything below it is of the morphologically precocial type. That the same value of 5 also separates the altricials from the precocials in birds reflects physiological relationships between brain size and maturity at birth that still need to be clarified in detail.

Since today, paleontology and comparative morphology allow us to make very concrete statements in many important areas of mammalian phylogeny, we are in a position to give the multiplier factors for certain ancestral groups with considerable certainty. This, in turn, permits us in many instances to provide the individual values for the different mammals with a tendency indicator (or "trend," to use a fashionable expression), thereby raising the informational value of the individual figure. For example, the figure for the brown hare is definitely related to the higher one for the rabbit and gives evidence of the evolutionary step from the altricial [primary ontogenetic type] to the precocial [secondary ontogenetic type] type. The figure for humans, however, is just as clearly to be

---

Figure 1.8. Baby rhinoceros on the day of its birth. After a gestation period of sixteen months (477 days on the average), the baby rhinoceros *(Rhinoceros unicornis)* is born; it weighs 60–70 kilograms and in outward appearance looks like a miniature of its mother. Even the folds in the skin were formed in the womb. However, the horn, from which the animal gets its name, does not appear until after birth. The large ungulates repeat during their fetal stage, as the human also does, the period of eye and ear closure, as though in preparation for the early birth of an altricial infant; millions of years ago, long before rhinoceroses existed, their ancestors, which were born in the altricial state, did the same. The first true ungulates in geologic history probably had gestation periods of 120 to 150 days. The different basic forms of horselike animals, ungulates, and elephants came first, appearing slowly, during the Cretaceous and Tertiary eras, from small ancestral forms, perhaps the size of a fox. We have many good reasons to assume that increased brain development did not begin until after the transformation in form had taken place. In earlier times, newborns must have gone on to increase their brain weight to about four times what it had been at birth, as is still true for young wild boars. It was not until the gestation period grew longer that ungulates were able to be born with about half of their final brain weight already developed. The picture shows the baby on the first day of its life, at its first nursing. The raising of Indian rhinoceroses in the Basel Zoological Gardens has provided important new data on the life of this large, endangered animal. Photograph: P. Steinemann, Zoological Garden, Basel.

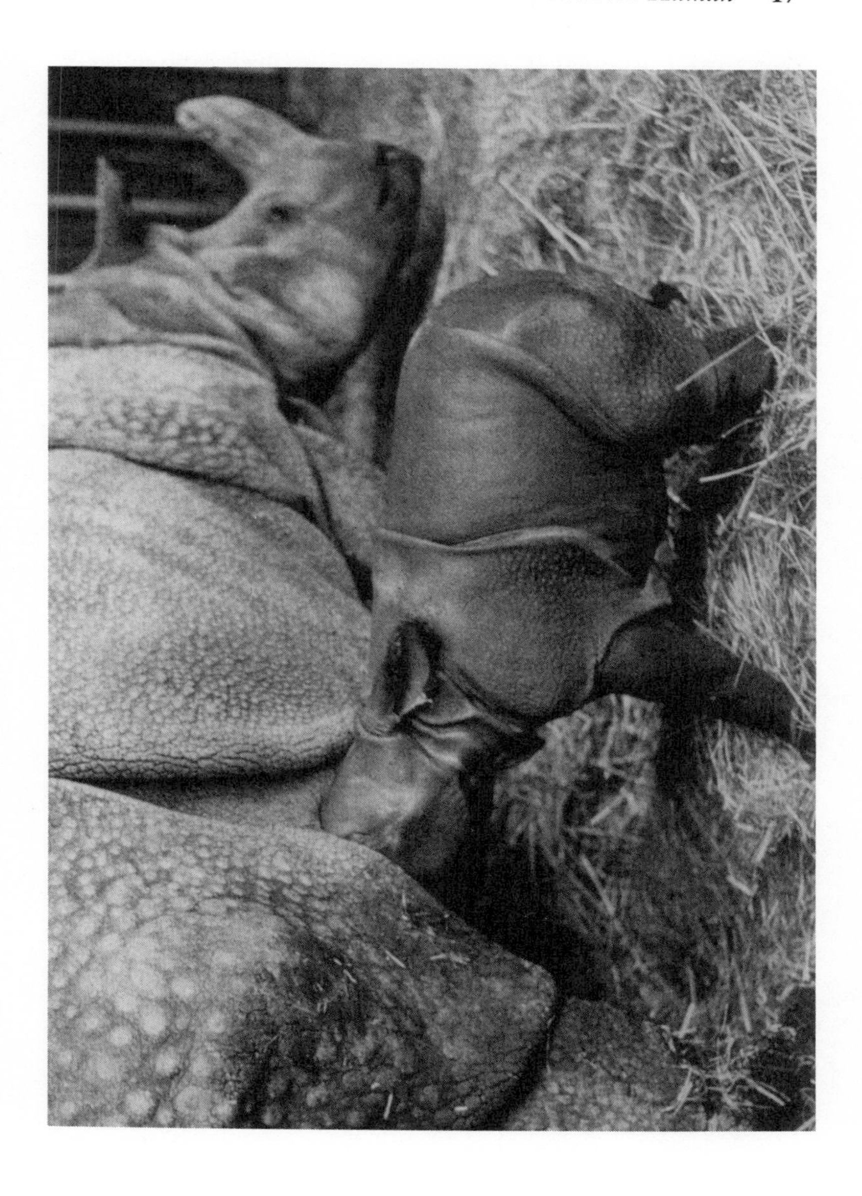

ranked with the lower figure for subhuman primates and indicates the path running in the opposite direction, from the secondary ontogenetic type to the new, secondarily altricial type typical of humans. The multiplier factor of wild boars, although very close to the figure valid for humans, has a different meaning with regard to evolutionary trend: it lies midway between primary and secondary ontogeny. Our table (table 1.6), much as it needs additional data, indicates important phylogenetic processes; there is, for example, the astonishing special instance of bears, in which there seems to be a regression within the primary altricial type. In this group, the formation of the anterior brain seems to be unusually delayed, as a comparison with other carnivores shows.

The evolutionary path of primate ontogeny is clear: from ancestors that belong to the group of insectivores with primary altriciality, development probably proceeds through types that were similar to the Tupaia and on to the rank of the precocial type, which characterizes the entire subhuman group to a striking degree. The separate path followed by humans is that of a "regression" to a new altricial form, and the figures we have thus far permit the formulation of well-supported statements on the subject of brain development at the hominization stage.

The multiplier factors for the great apes and for humans were taken from the research of A. Schultz (1956; see note 3 under Hofer). They are based on both brain weight and measurements of skull capacity. Many of these data should be supplemented with further material; for example, the figure for gorillas comes from a single bit of evidence from one newborn. At the time my "fragments" were first published, there was no hard evidence at all. My efforts to determine multiplier factors for the great apes were based on the oldest fetal and the youngest postpartum brain weights for all three apes and gave a first estimate of about 130 grams for the birth weight of all three. It has been shown, however, that this figure is too low and that a value of 180–200 grams is closer to the truth. For humans, a more accurate brain weight is 360 to 380 grams, on the average.

Of critical importance for our problem is the fact that for all the great apes, brain weights at birth are close to one another and that humans, with their much higher brain weight at birth, stand in contrast to this entire group. This corresponds to the total weight at birth: about 1,500 to 1,800 grams for all apes—1,650 grams on the average—and about 3,200 grams (somewhat more today) for humans. The ratio of brain mass to body mass at the moment of birth varies only slightly from the great apes to humans, from 1:8.6 to 1:11.5, thus showing an average of about 10, whereas the same ratio for mature forms, from 1:49 to 1:213 (and even more for gorillas) can vary.

There is only one way to interpret the body weight of newborns so that it makes sense and enables us to understand the differences among primates: the considerably increased body mass of newborn humans is an adjustment of the entire body to the size of the brain at birth, already larger in relation to the mature form than the brains of anthropoids. The conspicuously high birth weight of humans is correlated with the high initial weight of the brain; this, in turn, is clearly related to the special position of the human brain within the primate estate.

2
The First Year of Life

A HELPLESS altricial infant—that is how the newborn human looks to a zoologist. Are we aware that this fact transgresses the rule for mammals? For a moment, let us try to imagine how a human would have to be at birth if it were really subject to the same laws of development as the forms related to it. Such an attempt is not just idle speculation; it serves to establish a possible design against which the unusual aspects of our actual development can finally be measured. As we deliberate, we are only looking for a basis for comparison, for a model; it is not intended to represent the developmental state of any ancestral form whatever.

Physiological Early Birth

NEWBORNS OF all highly organized mammalian groups are precocial, and their sensory organs are well developed and capable of functioning. In form, apart from some slight proportional deviations, particularly in the size of the head, these newborns are miniature versions of the mature form, and their behavior and locomotion are to a large extent the same as their parents'. The infant also has command of the means of social communication that are typical for its species. This is the state at birth for ungulates, seals, and whales, as well as for anthropoids. As we have seen, with regard to development of overall form these statements are also true for the great apes, about which we will have more to say. In addition, many specialized rodents with a reduced number of young and (in the porcupine group) longer gestation periods, as well as the extremely specialized anteater and the South American sloth, which have only one offspring, follow the same laws.

In accordance with this definition, a true mammal of the human type would have to have a newborn whose bodily proportions are similar to those of the adult, one that can assume the erect posture appropriate to its species, and that has command of at least the rudiments of our communication system—language (and the language of gestures). This theoretically necessary stage does in fact exist during the course of our development: the stage is reached about a year after birth. After one year, the human attains the degree of formation in keeping with its species that a true mammal must have already realized by the time of its birth. Therefore, if the human were to arrive at this state in the true mammalian mode, our pregnancy would have to be longer than it is by about that one year; it would have to last for about twenty-one months. This figure of twenty-one months does not of course have to be an absolute figure. It depends a great deal on how closely we want our design to resemble the mature form; accordingly, we would have to "ask for" a few months more or less. Of critical importance for our further investigation is the necessity of stipulating a gestation period roughly one year longer for a manlike mammal: for a true animal-man *(Tiermensch)* or human-animal *(Menschentier)*.

In so doing, we are not indulging in impossible flights of fancy. Such long gestation periods do exist. The Indian elephant gives birth after twenty-one to twenty-two months, and this nimble baby elephant, about one meter high at the shoulder and weighing about 100 kilograms, is a good example of all the demands we have just spelled out. The gestation period of sperm whales is probably about sixteen months; this animal, too, gives birth to a well-developed baby: at birth, the "little one" is four meters long. Our preliminary conclusion is only that the actual length of human pregnancy is much less than it should be for typical mammalian development at our level of organization.

There is no serious disagreement about the statement that the human birth state is a kind of "physiological," or normalized, early birth, and the assertion in this form is also not new. But the contrast to the developmental norm for higher mammals has not been recognized: the actual problem posed by the length of pregnancy in humans has been blurred by the suggestive power of the commonalities that link humans and the great apes. What is lacking is a frame of reference that will enable us to see the peculiarities of our ontogeny more distinctly.

Many of the special aspects of the developmental processes that

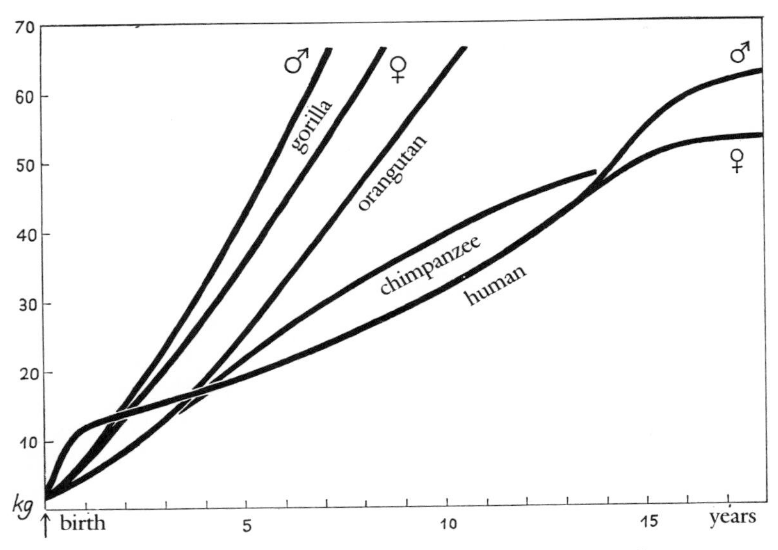

Figure 2.1. Weight increase in great apes and in humans. The curve for orangutans is based on data from G. Brandes; that of the chimpanzees, from Grether and Yerkes (the curve applies to a relatively lightweight race of chimpanzees). Once again, the gorilla curves reflect the measurements taken of the young born in Basel (I thank Dr. E. M. Lang, director of the Basel zoo, for the material). The divergent, differentiated growth mode of humans is clear: sharp growth during the first year; afterward, the curves for all the great apes intersect the much more slowly rising curve for humans, which goes through a new phase at puberty.

take place after a human is born, particularly the early period of postfetal growth, are easier to understand if we look at these events in the light of our conclusion. Human growth during the nursing period is characterized by the very intensive, early increase in mass: it contrasts sharply with the slowness of all later bodily formation, a fact that continues to be regarded as one of the human's most distinguishing characteristics.

There is a striking contrast between the early growth of the great apes and that of humans. The weight of the apes increases at a relatively even rate, beginning low and, by the end of the observation period, surpassing in all three species the body weight of humans of the same age. By contrast, our own growth curve begins far above the starting point for the apes, rises considerably during the first year, then makes a conspicuous transition to the special, very slow development that remains typical for us through the years. This sharp division of the growth curve into two parts during our first years is found in no other precocial type, not even in the great apes. It is peculiar to our species. As early as 1903, the pediatrician E. von Lange called attention to the fact that during our first-year growth in length, fetal proportions seem to be retained, and in 1922, R. E. Scammon again raised the issue of the fetal nature of this early period in regard to growth in mass.[9] Since that time, the fact has attracted little attention, for it seemed to be merely an isolated oddity. In our frame of reference, this fact acquires meaning. The strong growth appears particularly vividly if we do not follow a curve that is based on many averages and is therefore balanced, but look at a great many individual curves, which, of course, are harder to come by. In such curves, the weights between five and six months and between eight and nine months after birth are consistently above the points provided by averages.

The transition from the early type of growth, designated as fetal in many quarters independent of our own discussion, to the typical later growth mode takes place precisely at the time when the little human baby attains the "birth state" already defined for the typical higher mammal. We can probably advance this circumstance as a further argument in support of the view that there is a particular biological significance in the moment when such important events occur. At the end of the first year of life, the moment comes that must be considered as the time of parturition for any true mammal of humanlike organization.

A Chapter on Comparative Evolution

PERHAPS THE special nature of the timing of our actual birth will appear more clearly if we extend our comparison beyond the mammalian group and attempt to gain a more comprehensive view of the ontogenetic situation in the higher vertebrates. The concept of the course of human development presented here has also gradually arisen from research carried out pursuant to developing such an overview; consequently, I hope what follows may demonstrate how the zoologist is led from comparative studies to problems of anthropology.

Every comparison of the developmental course of warm-blooded animals must begin with the reptilian state. We derive this certainty first from the reptiles' position, granted to them by the theory of evolution, as the ancestor of birds and mammals. This concept is in turn based on morphological research that demonstrates the fact that in many respects, the level of organization in reptiles is relatively simple. Of greatest importance seems to be the low degree of brain formation. Morphological facts are, and remain, of primary significance; evolutionary consequences must under every circumstance be derived from them.

Individual development in reptiles is distinguished by a juvenile state in which, upon leaving the chorion behind, the animal already resembles the adult to a large extent: a miniature version of the adult hatches from the egg. Consequently, most young reptiles are completely self-sufficient, and in the infrequent instances when the young remain for a time with the mother, the role of the parents is limited to leading the way and protection. In most instances, however, the young generation never sees its parents at all. The hatching sea turtle is guided from the beach to the sea by its own instincts; the young adder, like the lizard, goes its own way right from the start. All these young animals must be termed true precocials. Only a few modern reptiles exhibit actual brood care, and even in those that do, contact between old and young animals is slight.

Birds exhibit a developmental mode that is very similar to that of reptiles. An extreme, particularly striking example of self-sufficiency in juvenile birds is provided by the megapodes, a group of birds found in the Austro-Malaysian realm. But other fowls, too, as well as ostriches, plovers, ducks, and geese come close to matching this extreme. In contrast, many other groups of birds are characterized by the striking dependency of the young birds on the adults. The secondary nature of this dependent juvenile existence makes itself known through the fact that it is very often connected to a reduction in the number of offspring to a single one, to one egg per clutch. The highest ranking birds—owls, parrots, woodpeckers, and the passerines—all exhibit the helpless, dependent altricial state. High levels of brain formation require a long period of development, during which the neuromuscular system cannot function as it will when it is mature. This predicament, the long period of inability to lead a self-sufficient existence, is compensated for by the sharp limitations imposed on the lives of both the mature and the juvenile birds: the parents carry out the functions that the young, immature bird is still incapable of.

A very similar solution has been arrived at by many mammalian groups. In the material that follows, we shall refer mainly to the low organizational levels of the so-called eutherians, since our own form and the primates belong to this group. Marsupials would present a similar situation, but they pose additional special problems that lead us away from man. We shall therefore stay mainly with insectivores, many rodents, and small carnivores. All of these exhibit, as has already been mentioned, the altricial state. The high-ranking mammals stand in contrast to this group in the same way that separates the low-ranking birds from the high-ranking ones: the increased level of organization of the nerve apparatus in turn extends the period of time during which the nervous system is unable to function as it will when mature.

How, then, do the more highly organized mammals cope with this extended period of inability to lead a self-sufficient life? Conceivably, it might be through more intensive brood care: a lengthening and intensification of the altricial state. However, the reality of life contradicts the conceivable. The solution for high-ranking mammals is totally different: a considerably lengthened fetal period within the mother's body, one so extended that the state at birth in such mammalian groups, as we have already seen, represents a new kind of precocial mode, one in which the infant seems still bound to the parents only through its need for milk—a "dependent precociality," in contrast to the situation, so similar in many other respects, for reptiles.

The singularity of this type of precociality when compared with the similar state in reptiles is also reflected in the fact that higher mammals, almost without exception, have an extremely reduced number of young, whereas reptiles always have many! Often, the number of offspring is also an indication of rank.

Once again, we encounter in the overall picture of warm-blooded animals a significant jump in the level of organization: within the primate group, in the contrast between the great apes and humans. Once again the developmental span, from the egg to a level of environmental adaptation, particularly with respect to behavior, that is in keeping with the mature form, grows longer. Again the question arises of how animals cope with this period of extended dependency, which is part of their development. We know the solution typical of the higher mammals; as we have already seen, it would be another extension of the gestation period by almost a year. But once again, as in the abrupt change between lower-ranking and higher-ranking mammals, we find in reality not the solution that comes readily to mind by making comparisons, but a new and unexpected one!

In humans, the extension of the dependent developmental period required by increased brain development is handled in a manner similar to that found at many simpler levels of organizations. The path taken by high-ranking birds is preferred to the lengthening of the gestation period, and a period of intensive parental care and protection is introduced. The newborn human, basically a precocial type, ends up in a special kind of dependency, which is why we have termed it "secondarily altricial" (Portmann 1942). With this special kind of dependency, the human stands alone among the mammalian groups. We should not let ourselves be deceived by the seemingly impressive relationship between the great ape mother and child as displayed through gesture and behavior, for in the well-protected, newborn ape no essentially new possibilities for behavior and movement or for means of communication arise, whereas in humans, essential developmental stages occur during this special extrauterine time (and in fact, at its very end!). In apes, the final states, already present and fully formed, mature in rapid succession. In humans, however, it is only at the end of the period that the really species-specific behavior, movement, and speech appear. This observation by no means minimizes the significance of the sheltered infancy of higher animals, in particular of monkeys, apes, and the great apes; the juvenile period includes tradition, but does not lead to a historical mode of life. We shall have to return to the parallels and contrasts with regard to the highest primates.

When we say more simply that the human is born almost a year "too soon," it is our intention to stand firmly by this purely comparative statement and to pose no further questions about possible origins of this special state.

Figure 2.2. African elephant on the day of its birth. Assessment of our own state at birth must be oriented around the states of the other groups of highly developed mammals, all of which bear well-developed young of the altricial type: ungulates, whales, seals, apes. Increased bodily size, the higher level of relationship with the world, in particular of social relationships, enable the species to be preserved with one (rarely two) young per litter, with longer periods of gestation and nursing. Elephants demonstrate the extreme of this level: after twenty to twenty-two months of development in the womb, the young is born, measuring about one meter high at the shoulder and weighing 90–100 kilograms, and it is very soon able to follow its mother and move with the herd. Our picture shows the first African elephant *(Loxodonta africana)* to be born in Basel, on the day of its birth. It has only been thanks to observations made in zoos during recent decades that we have acquired more exact data on the growing up and maturing of many mammals that we were familiar with in form but about whose way of life we very often knew nothing. Photograph: P. Steinemann, Zoological Garden, Basel.

Figure 2.3. Gorilla. The life of the great apes has at long last been observed in detail only in the last decades, more particularly, since 1950. Only now do we recognize their unusual nature as special forms of a higher mammalian group. Formerly, they were usually thought of as curious relatives of our own life form and evaluated differently, according to the spirit of the time: as "men of the forest," closely related to us, in the seventeenth and eighteenth centuries; and as wild beasts a hundred years later, when gorillas roused the imagination to particularly brutal representations. Observations carried out by the American Karl Akely after 1923, and, more recently, the detailed studies of his countryman George Schaller (1963) provide us with a completely new picture of this mighty quadruped, a picture that will probably slowly replace the old notions. The peaceful way of life of this huge ape shows us the same thing we see in chimpanzees: a rich social existence, which has, however, developed its own special form after so many millions of years that we cannot say what possibly corresponded to our own primitive form. Photograph: P. Steinemann, Zoological Garden, Basel.

Assessment of the Duration of Pregnancy

THE POINT on the developmental continuum at which birth occurs can be commented on from very different points of view. Either all the facts of the life-form of a species are taken into consideration, or the evaluation of the timing is based more on the circumstances immediately surrounding the moment of birth. The latter approach dominates most discussions of this problem.

Thus far, attempts at explanation have been based on the corresponding durations of anthropoid gestation and human pregnancy, a similarity that allowed a closer parallel to be assumed than appears to be true today based on data from chimpanzees. The duration of the gestation period seemed to be a very similar, genetically determined event for all higher primates. Thus, Grosser (1941; see note 10) explained the occurrence of birth as being occasioned by the genetically determined lifespan of the fetal placenta. We do not believe, however, that the idea of looking for factors affecting birth in the genetic determination of the lifespan of placental cells is a valid one, even though we have a high regard for the significance of these genetic factors.

Biologists have drawn other, more complex correlations into the consideration of the issue. Thus, we often see the moment of birth as being connected to the high level of human brain formation, as being "determined" by the dimensional relationship between the width of the birth canal and the circumference of the baby's head. There is no doubt that such a relationship based on size exists. But it is impossible for the biologist to orient this relationship with regard to cause and effect: to say, for example, that the adult form is the given and that therefore, the size of the mother's pelvis determines the possible mass of the infant's head. Arguments of this kind are probably used more or less consciously all too frequently in biology, but no matter how often they are used, they do not become more acceptable. In such an instance, it is much too simple to explain two sizes that occur in relationship to each other by saying that one is a given and the other is dependent on it. Not until puberty does the female pelvis, influenced by sex hormones, take on its typical form, one that is different from that of the male; nor should this pelvic form simply be taken as an established given. At any rate, since the form of the pelvis seems to be tuned to the special requirements of viviparous (bearing live young) development, there is no possibility of proving that the dimensions of the birth canal have reached a threshold, thereby determining the size of the infant's head. No one knows whether the human organism is capable of forming a larger birth canal, one that would be in harmony with an infant head of larger diameter. Let us not forget that in precocials, as term approaches, the shape of the pelvis changes to accommodate the birth and then reverts to its former state. All that can be established scientifically is the fact that the shaping of the female pelvis and the formation of the child show a correlation of size. The question remains open, however, whether the same correlations might also occur among other dimensions in close proximity to the two related elements observed.[10]

The factors at work in fetal development, like those that direct the functions of the mature organism, are components, and only components, of one structure, and all of them are part of the one, genetically determined basic structural design of that organism. This primary fact must be taken into consideration in any investigation of physiological or evolutionary relationships. Consequently, we cannot regard any factor formerly held responsible for the temporal determination of the moment in human ontogeny when birth occurs as being the one dominant element. Even considering, for example, the existence of a certain heirarchy of hormonal effects, significant as they may be for practical intervention in the course of events, the biologist must still recognize the final systematic assembly of even this heirarchy as simply a part of the basic structural design. Even the leading organ in such a system, the hypophysis, for example, is itself only a "component"; it is a "leader that is led," not the "court of final appeal."

The comparative investigation of developmental relationships in mammals leads us to assume that the length of human pregnancy does not correspond to the complete gestation period that would be suitable for a mammal of the human's high level of organization. On the other hand, looking at the problem from a physiological point of view does not result in any sound reason explaining this early moment of birth, atypical for mammals, as a "technical" necessity. Therefore, it seems justified to investigate anew the position of the moment of birth, taking into consideration a broader range of facts about the human life-form in hopes that, when reviewed within a comprehensive field of observation, the meaning of this "physiological early birth" will be seen more clearly.

3
The Human Life-Form

HOW LARGE a frame must we use to display human development in all its singularity?

Scientific attempts to interpret this singularity, to the extent that biologists are even aware of it, all proceed from the concept, often only a resonant undertone, that the distinct species man arose "very gradually" from animals as a late phase of a mammalian evolution. "Very gradually": the magic words of evolutionary theory, by means of which even the most widely differing elements can be traced back to a common origin if only we devise enough small, transformational steps and add the required incalculable eons that geological research says we are entitled to!

Life-Form and Course of Development

EVOLUTIONARY THEORY also sees a gradual rising from the animal state to the human in our individual developmental history: it even sees in the process by which an individual comes into being a reflection of the development of the human form over geologic time. According to this widely held view, the earlier the developmental phase observed, the less it contains of that admixture that finally makes a human out of an animal's body, and therefore the more exclusively is it an object of biological research. Development that takes place before birth, especially its early stages, is considered to be particularly animalistic. In keeping with this line of thought, the significant changes that take place toward the end of the first year of postpartum life are designated by many biologists and psychologists as nothing less than the actual onset of becoming human. This view has been expressed even more graphically: that the end of the first year is the time when man passes beyond the chimpanzee state! It is not without reason that the study of early development is actually an area of research within zoology, and that interpretation of that stage from the point of view of the theory of descent appeared to many as zoology's real contribution toward making anthropology a science.

Our description has indeed emphasized the human singularity of our infants to such an extent that we should now venture to say that the just-mentioned zoological interpretation of early human development is inadequate and in many respects misleading. The special nature of infants is the result of an independent human type of development before birth. Even the early prenatal development is the "ontogeny of a human," not a kind of schematic primate formation in which the stages of animal systems appear in sequence, as in a graduated classification. If in many investigations it is justifiable, even essential, to emphasize general traits of a single developmental stage provisionally, or even, in order to find certain general laws, to observe them exclusively, it is just as important for insight into human development to stress additionally the special nature of this developmental path at all stages and as much as possible. The schematic similarity of the early developmental phases conceals the secret difference, just as the simple spherical shape of many egg cells leads us to assume that they are identical, whereas experimentation reveals hidden, extremely different potentials.

The framework within which we observe human ontogeny is, therefore, not the one that evolutionary theory usually uses for mammalian development; rather, for us, development is part of a larger, more comprehensive fact of human existence—as in those old paintings where different episodes of one life are depicted as taking place all at once, each having consequences for the whole.

The human form of existence—for the natural scientist, this

cannot be simply the visible form and the functioning of its elements. Not until behavior is also taken into consideration will the true extent of human singularity be revealed; only when we take this aspect of our being into account will the field for a comparative-biological evaluation, for a more comprehensive understanding of human development, be found. Since, however, ontogeny is the methodical, orderly development of a complete form of existence, the separate steps can only be comprehended in their true light as acts in this special drama. Even the most rigorous, cause-oriented research into a developmental course will not be eroded by the fact that this sequence of cause and effect is directed toward a goal that is known to us.

It goes without saying that for every embryologist doing morphological research, a prerequisite for the interpretation of developmental states is a precise knowledge of the final form; it is based on this knowledge that we understand the developmental steps. It is necessary to make this very trivial assertion, for the same principle has a much broader application; and yet, very little attention is paid to it outside the narrow circle of embryology. Even the curious nature of behavior—of the entire mode of life of the mature form—has its appropriate correlations fitted into the peculiarities of the developmental plan.

Therefore, we must also find in our development traces of the origin of the special mode of behavior that is peculiar to us humans. As the morphologist studies the mature form in order to explain the germ, so must we intergrate what is typical for our way of life into the sphere of biological research.

We place this analysis of our behavior at the center of our deliberations. Not that we would slight the somatic peculiarity of our bodily structure—on the contrary—but, currently, a shift in emphasis is being felt. Somatic characters have been the center of the attempts at explanation long enough; this approach has not been successful in establishing a comprehensive picture of human development. We hope to show that the change in the manner of observation is more than a heuristic attempt and that as behavior is investigated, many essential traits of our bodily structure and their mode of origin will finally be incorporated into a larger whole.

Those who, from inner predisposition, from the need for graphic results, have seen their mission to be research into visible phenomena will turn to describing behavior only with reluctance and much caution. Yet it is exactly the scientific analysis of form that compels such overstepping of the boundaries separating the traditional areas of study, for every attempt to position the human phenomenon within the realm of forms leads back to that gate through which one enters the realm of the mind: a stranger realm, but the only promise of a new horizon and a way to survey it. The attempt must be ventured. If, in the future, better insight continues to work with the same love of subject matter, much that is unsatisfactory in the views attempted here will probably be ultimately replaced with enduring truth.

One should not look for a closed system in the following points on human peculiarity; they are far too conditioned by feelings of the magnitude of the object to be described. A closed system would imply the presumption that we have completely understood the innermost nature of all of human behavior.

Only features will be given that are particularly significant for our perception of the human-animal relationship. We are not able to arrange these characteristics in any kind of order, for, in actuality, they are given to us in an intertwined complexity. Any one of them might be placed at the beginning; the sequence chosen here represents our attempt at organization and is only one of the possible ways of presentation.

Central Nervous System and Life-Form

OUR POINT of departure is the striking difference between human and animal in the importance of instinct. Behavioral research correctly refers to the great significance of tradition, of how modes of behavior are imitated and assimilated in the higher forms of birds and mammals. As we assess the period of childhood, we shall also take up the role played by tradition in the behavior of our nearest relatives, the anthropoid apes. Nevertheless, it is true that chimpanzees and gorillas raised by humans do not become human, but most follow in decisive steps the relationship to the world marked out by the genes of their species. Whereas, in animals, essential modes of behavior are determined by that organic precondition we call instinct, in humans even the most instinctual part of behavior—sexuality—is open to the extensive freedom of personal decision; even in this area, one for which we too have a powerful instinctual basis, the possibility for sharp conflict between extremely different kinds of behavior exists.

This possibility for free resolve is even more significant in other areas of life that are not so closely tied up with preservation of the species. We want to guard against making a superficial attempt to localize the just-emphasized difference between human and animal neurologically by relegating it to organizational correlations. And yet, we must keep in mind many facts made available by research into the central nervous system, facts that are important for a comprehensive understanding of the difference. Thus, we should pay attention to the fact that the centers of the hypothalamic region of the midbrain in lower mammals are of a strikingly more complex organization than those in anthropoid apes and humans. Even in monkeys, those centers are higher in this regard than the corresponding regions in the brains of the great apes.[11] This morphological finding must be understood in connection with a diminution of the power of instinct and the shift of the centers of important functions to the cerebral cortex. These circumstances make the enormous increase in size of the human cerebral cortex and its functions even more significant. Nevertheless, we do not want to underestimate the instinct in humans: the hormonal glands add considerable weight on the side of instinct, for their organization is to a large extent in accord with that of lower mammals.

The human's relative weakness in instinct is counterbalanced by a powerful increase in other, more central drives. This increase is expressed first in the enlargement of the hemispheres, in the increase in mass of all of those structural elements that are particularly significant for the higher vertebrates.

At the very beginning of this investigation, the attempt to arrange ontogenetic types in a particular ranked sequence posed the problem of giving a relatively sure basis for this ranking. We had to examine in detail the various methods already in existence so that, using the formation of the highest organ of relationship to the world and of coordination of the organs—the brain—we could work toward finding an order; we came up against the problem of cerebralization.

We would have to digress considerably to give the background of these attempts. However, it is necessary for our investigation that we give a brief explanation of the significance of the index figures that we compiled for birds in 1946–47 and that K. Mangold-Wirz (1950) compiled for mammals.[12]

We compared the mass of a part of the brain that predominantly serves the basic functions of life (the so-called brain stem) with the mass of a part of the brain that mediates with the environment (the cerebrum), or that regulates the finest interplay of our parts (the cerebellum). The comparison results in a quotient that reflects by how many times the mass of the higher nerve center exceeds that of the lower one. Since in higher brain development, the brainstem itself also evolves, we choose as a measure for our indices the lowest base value for a particular body size.

For our investigation, the index for the anterior brain, or cerebrum, is the most important value. First, using a few simple comparisons, we must test what such figures express. Let us first look at the cerebral indices in groups of mammals that have similar body masses. The contrasts between the indices are definitely independent of body size. A few examples appear in table 3.1.

Next, we shall compare mammals that are closely related but

that differ in body size. If, in such instances, the indices are alike or similar, it shows that this figure is not directly dependent on body size but expresses other facts of organization. Three groups of such contrasts illustrate the usefulness of the index figures (table 3.2). Moreover, even before we took up mammals, we tested the validity of the information provided by our indices for birds against a very large body of evidence.

Particularly enlightening is a glance at the indices of the most

TABLE 3.1.

Body wgt. 1 kg		*Body wgt. 3 kg*	
hedgehog	0.77	armadillo	2.8
muskrat	2.75	marmot	4.3
wild rabbit	4.6	common hare	5.1
European polecat	12.90	cat	12.3
lemur	13.1	macaque	32.4
Body wgt. 50 kg		***Body wgt. 70 kg***	
jaguar	19.1	spotted hyena	16.9
aoudad	21.3	guanaco	21.7
chimpanzee	49.0	reindeer	24.0
		human	170.0

TABLE 3.2. Indices for the Cerebrum

	Body wgt.	*Index*
shrew	17 g	0.76
hedgehog	928 g	0.77
leopard	53 kg	18.3
tiger	182 kg	18.7
black buck	29 kg	17.8
yak	250 kg	19.5

important part of the primate brain. It is a great advantage of our index method that it gives several index values for one brain. The result is *index groups,* which enable a nuanced characterization to be made. Table 3.3 and all other numbers presented here have been taken from studies carried out by Katharina Wirz (1950).

First of all, these numbers show the anticipated rise in the indices of the cerebrum, the cerebellum, and the brain stem in the series that runs from lemurs through Old World monkeys and great apes to humans. Furthermore, they illustrate the lowering of the olfactory lobe index, which runs as expected if we disregard the surprising peculiarity of chimpanzees, whose olfactory lobe index lies far below that of the human. We note above all the distinct singularity of the New World monkeys, which are represented by the capuchin (Cebus). All of its indices are higher than those of the chimpanzee; only the index of the cerebellum is similar to that of the great apes. These details offer impressive testimony to the special position occupied by South American monkeys. On the other hand, the comparison between baboons and chimpanzees is a good one to clarify the position of the great apes. It shows first that in baboons, the brain stem index is greater, and in chimpanzees, on the contrary, that the two integrative areas of the

TABLE 3.3.

	INDICES OF					
	Cerebrum	*Olfactory Lobes*	*Cere-bellum*	*Brain-stem*	*Body Weight in kg*	*Brain Weight in g*
lemur	13.5	0.98	3.24	3.72	1.5	23
capuchin monkey	53.7	0.27	7.25	6.92	1.8	69
guenon	33.9	0.42	4.90	4.79	2.8	56
baboon	47.9	0.33	5.13	6.24	12.5	170
chimpanzee	49.0	0.03	7.59	4.47	61.0	363
human	170.0	0.23	25.70	10.0	70.0	1,290

cerebellum and the cerebrum are relatively larger. The same comparison makes it clear, however, that the position of the great apes is not simply intermediate between man and the lower apes, but that the interval between the ape level and our hominid organization is quite large.

I emphasize this interval in particular because it is apt for influencing ideas of anthropogenesis. Contrary to the concept that the chimpanzee is very close to the ancestral line of hominids, as many anthropologists were maintaining not so very long ago, a different opinion is now widespread: that very early, in the Oligocene or Miocene, hominids, branching off from ape groups with smaller bodies and higher cerebralization, began to develop separately. The index figures of the capuchin could support the concept of such a primitive type, which would not need to have had in addition the characteristics of the New World monkey.

The index formula for the human brain shows that we should not only pay attention to the cerebrum, but that both the brainstem and the cerebellum have undergone very large increases above the level found in the other primates. The brainstem index of 10, typical for humans, says that the primitive part of our brain is ten times heavier than the same part of the brain of one of the lowest mammals of the same weight, an example being the giant kangaroo! Within this brainstem mass are the portions of the pyramidal tract that lead from the cerebrum to the spinal cord and that have a large share in the centralization of neuromuscular potential.

The calculation of brain indices should not lead to a purely quantitative consideration of the brain. Nor should the comparative values given be used now in premature attempts to decide among the contradictory viewpoints on the evolution of hominids. We see the current significance of the calculation of such indices primarily in the fact that they demonstrate the singularity of our life-form, thereby reminding us of the true extent of the problems facing research into evolution.

Another table (table 3.4) shows how the different types of

TABLE 3.4. Indices of the Mammalian Cerebrum (after K. Wirz 1950)

	<2.5	*2.5*	*5*	*10*	*15*	*20*	*30*	*40*	*>50*
shrew	0.7								
hedgehog	0.7								
mole	1.1								
microchiropteran	1.1								
hamster	1.8								
mouse, rat	1.9								
squirrel			6.1						
porcupine			7.9						
agouti			7.4						
rabbit		4.6							
hare			5.1						
armadillo		2.8							
marten				13.2					
dog					16.7				
hyena					16.9				
cat					18.4				
bear						23.3			
dolphin									119
hyrax			8.5						
elephant									104
tapir				12.6					
horse							33.3		
hog				14.1					
camel						27.6			
deer						28.2			
giraffe						29.5			
horned animals						20.1			
prosimian				13.5					
New World ape									53.7
guenon							33.9		
baboon								47.9	
chimpanzee									53.0
human									170.0

mammals line up in our attempt to find an objective understanding of brain size. From top to bottom, the table seeks to show, as far as is possible in a linear arrangement, the increasing significance of the cerebrum. The increase in brain formation is very clear within individual related groups (rodents, lagomorphs, carnivores, primates). The significant interval in a few broad related groups is impressively demonstrated (hyrax/elephant in the group Subungulata; tapir/horse in the odd-toed ungulates). The prominent position of dolphins and elephants stands out clearly.

The mammalian indices extend in general from 0.7 to about 50 for the larger groups. Only three types exceed this limit, and the great apes are not one of them; rather, it is the elephants and the dolphins that, along with humans, make up the three. Each of these three peaks poses a special question and reminds us that the numerical values can have quite different meanings, and that they offer nothing more or less than a first, general orientation. It should be emphasized once again that, owing to the manner in which they were arrived at, the index values are to a large extent independent of body size and therefore in many respects enable comparisons to be made. The values for dolphins are particularly interesting, for the dolphin brain has long been compared with that of the human, both in size and also in many structural details.

Our figures are objective values for the relative size of brain mass, especially of the cerebrum. They are never a measure of mental capacity, of intelligence; what they mostly do is pose new problems.

Thus, the difference between our brain and those of the related great apes rests not just in the contrast of index values. Nevertheless, the numerical difference provides an impressive measure of the changes that have led from the subhuman primate level to the special relationship humans have with the world.

Psychologists and evolutionary scientists must continually keep in mind the enormous increase in simple mass of the anatomically comprehensible substrates of essential mental processes in man in comparison with most higher mammals. The powerful increase in mass of the cerebral cortex and its tracts is related to the diminution of instinct. This contrast allows the peculiarity of the instinctual drives that appear to be allotted to the human way of life to stand out.

Even when we relate the increased cortical mass in humans to the special instinctual drives that impel our actions, we still have not the slightest knowledge of the physical organization of these drives. We can understand them only through the ways in which they express themselves, and call them, for example, the human "will." Every search for a comparison that would relate this phenomenon of the "will" to other natural phenomena immediately leads far beyond the realm of nervous processes into the area of the influences we know of operating in the growth of organisms and in their regulated developmental transformations. Nowhere do we find a more certain basis for comparison, and so our search for analogies dissipates as it flows into the inaccessible mystery of life itself.

It is not part of the plan of this investigation to seek in animals the "predisposition" that might be compared with the singular human quality of the will. For us, the very particular strength of the special human system of drives and its undisputed superiority in comparison with instinct-directed animal behavior is the focus of consideration. In assessing the singular nature of human drives, we should not underrate the hormonal component. Once again, as already pointed out with regard to the hormonal glands in general, the physiological basis resembles the general situation found in mammals. But the rhythmic activity of hormonal drives so characteristic of animal life, the regular alternation of the rutting season with periods of sexual indifference, is much more relaxed even in the higher primates, and has been practically done away with in man.

The lasting effect of the sexual components, the most conspicuous of these hormonal effects, leads to a very special result in our lives: to a continuous, lasting sexualization of many human drives;

but also to a significant permeation of sexual activity with the other motives for human behavior, which are always in force. No one can mistake the extent of the intensity and particular coloring of the entire human experience brought about by the continuous effect of sexual factors. Often enough, these effects, exaggerated and distorted, are granted absolute dominance, but more attention should also be paid to the continuous repression of the sexual drive resulting from the fact that it is always operating simultaneously with the other components of our guidance system.

It is in keeping with the relatively low level of development of prefigured, instinctive modes of behavior in man that no particular environment, no particular sector of nature seems to have been allocated as our habitat. There is no environment for humans that can be posited the way it usually can for an animal: the steppes or the forest, for example, rivers or mountains, or even the much more circumscribed territory of forest canopy, thicket, or rocky bottom. On the contrary, it suits our whole mode of existence for us to create a special "world" in any natural area we visit, to build it up out of natural components transformed by human activity. In contrast to the changes that an animal, too, can make in its surroundings—always just a corner of the assigned "environment" that has been altered by essentially instinctive activity—human intervention is freely determined, takes place anywhere that is accessible, and always includes realms inaccessible to the senses, consciously related to the past and the future. Just how significant for all of human existence is this continual going beyond what is accessible to the senses is vividly shown in the ever-present fear of demons held by many so-called primitive peoples. Many theories of descent would do well to consider how little such behavior is animalistic, how little "primitive." Attempts to derive the human mind from animal states always take particular pains to point out the origin of intelligence, the tool of thought. How little these theories are concerned with the rich proliferation of fantastic concepts created by the all-powerful, scarcely conscious mental life of the "primitives"! If, in evolutionary thought, we term this area of the mind "prelogical," the difficult task of deriving this mental state, which precedes the logical state—this "primitive mentality" —from the animal state rises up to face evolutionary theory. Many scientists who work on theories of descent begin to suspect the vastness of the area that remains to be studied only when faced with this task. It will be a good thing if we first eliminate the term "prelogical," as is being increasingly done today, thus leaving the field free for objective research.

The special human world that is thought of as "culture" can be set up in opposition to Nature; it can be regarded as "artificial" in comparison with the natural environment of animals. There are all kinds of expressions for this rich, meaningful set of circumstances. This "cultural life" is such a general aspect of humanity that we find no human groups that are unsophisticated "children of nature" *(Naturmenschen)* in the literal sense, just as we know of no *"Naturfolk,"* for even culture in the most general sense of the word is a part of the behavior of all, even the most "primitive" peoples. Admittedly, *Naturmensch* sleeps in our souls, an image in many dreams; as a figure in contrast to ourselves, as an ideal, it is always ready to come forth. And there are many times when this image appears aggressively as a reaction against the ineluctable power and force with which the culture surrounds us. Then, sometimes, from this inner struggle the image of the noble savage—a better man than we, the civilized ones—emerges.

If we say that animal behavior is "tied to the environment" *(umweltgebunden),* we must call human behavior "open to the world" *(weltoffen),* meaning, with this wonderful expression, a treasure trove of creative behavior, which the individual can put to more or less worthy use; and, like an estate, it can also be squandered.

Objective Behavior

FOR AN animal, whether and to what extent a given component of nature affects its being, whether and how much the component has a vital meaning within the environment, is decided ahead of time through the animal's environmentally bound organization. In contrast, human design, open to the world, creates a completely different relationship to the natural environment. Even the most unlikely component of the environment can be meaningful to us; we are capable of separating any detail at will from the undifferentiated field of perception, and emphasizing it. Everything in the environment can pertain to us; even remote things can acquire meaning, just as that which is hidden from our ordinary senses can. And so, research is continually seeking unknown, prospective vehicles for meaning, whereas the most active, inexhaustible, sniffing, scenting, tracking animal is always seeking only the vehicles predetermined by its organization to be meaningful.

Is there anything in animals that compares with this human capacity for lending meaning to the meaningless and for the creation of new vehicles of meaning? In fact, there are in higher forms of animal life beginnings that hint at the possibility of "interest in the unfamiliar." The highest modes of behavior are found in species that are singled out, with good reason, as being "curious." In the training of animals, substantial results are obtained by making use of such curiosity.

Moreover, there are also curious, sporadic suggestions of the behavior emphasized here as being human in the play of the young of higher mammals—and only there—during a particularly rich phase of their existence, the period when these animals, and not without reason, seem so familiar to us, so close, and to be so related to us.[13] How to evaluate such traces, such scarcely noticeable suggestions of the human species, is not easy to decide.

One peculiarity of our open-to-the-world behavior is the remarkable capacity we have labeled "representation" *(Repräsentation),* which includes not only the continuous, many-faceted supplementation of what appears only one-sided and partial in the field of perception, but also the constant, additional imagining of things that are not present to the senses at all, of things that are distant in either time or space. Whatever beginnings of higher representation there may be in animals, they can never be compared to the extent of this capacity in humans. That we humans can experience an entity as a "torso," as something that needs supplementation and that will be supplemented without more ado through our inner activity, demonstrates the possibilities granted us by this capacity for representation. What a mighty intensification of the power of human expression lies in this mysterious ability to supplement or complete; what an event—the highest form of collaboration—when the artist involves us in his creation by consciously setting before us only fragments, which we, spectators and partners in the act, are to complete.

There remains to be pointed out a significant difference between our form of existence and that of animals: the possibility of conceiving of ourselves as an object completed by representation and of confronting this object as something alien. We can observe ourselves as well as other objects and events from a standpoint chosen, as it were, outside ourselves. All sorts of words have been used to try to describe and define this peculiar human existential possibility. For example, the situation of constraint imposed on an animal by its mode of existence is termed "concentric," to distinguish it from the possibility we have of stepping outside ourselves, which is termed "excentric." The same elements are expressed when animal behavior is said to be purely "subjective," and that of man is characterized by the additional capacity for "objectivity." The contrast is also expressed when we say the animal "lives" its life, whereas man "leads" an existence. We shall let the matter rest with such general statements, for the subject is a familiar one, and we wanted only to emphasize once again its full significance for our further investigations.

The contrasts just mentioned are so evident in animals' outward expression of their inner life, they confront us so objectively in verifiable behavior, that the inaccessibility of animal experience should not keep us from accepting the thoroughgoing difference that appears in our comparison. We know of nothing in the behavior of animals that would correspond to being devoted to a cause to the point of sacrificing one's own life, as is possible for humans for the most various of reasons. The high capacity for selfless love and devotion is given only to humans—however much this greatest of possibilities eludes many people all their lives and even the best of us sometimes.

Constrained by environment and protected by instinct: simply and briefly, that is how we can describe the behavior of animals. In contrast, human behavior may be termed open to the world and possessed of freedom of choice. In making these definitions, we want to emphasize positive aspects of a powerful, many-sided reality that has also been evaluated differently. Often, the negative aspect of the same subject has been overemphasized; in those theories, we encounter man as a being without instincts, as one that has been cast out of the secure haven of animalian environmental constraints. Sometimes, he seems in these accounts just like a convict who has been let out of the security of prison to fend for himself on the streets of life.[14]

4
*The Extrauterine Spring**

OUR ATTEMPT to characterize human behavior seeks to see both the animal and the human forms of existence as comprehensively as possible. The results of this overview should equip us to present the developmental modes of mammals and humans in relation to the kinds of behavior exhibited by their mature forms. We hope that ontogenetic facts heretofore unnoticed or unexplained will find their meaningful place within this expanded field of consideration.

The Typical Development of Higher Mammals

LET US first try to understand the development of a high-ranking mammal in relation to its environmentally constrained, instinctually assured life. The most conspicuous characteristic of this development is the growing up and maturing of the whole organic structure within the protection of the mother's body. All higher mammals have long gestation periods, the length of which stands in clear relation to the level of organization of the central nervous system. Following the developmental plan fixed in the genes of the species, the entire motor apparatus, the species-specific posture, and the typical system of instincts form inside the mother's body, in the uterus—all in harmony with an environment that has also

*Chapter title: the German reads "Das extrauterine Frühjahr," the last word referring to the season of new growth.——Trans.

been to a large extent genetically assigned to each species. The basics of movement and behavior take shape within the mother's body, far from later sources of stimulation and yet related to these future stimuli, to the environment yet to come. There is no difference between the appropriate formation of the hooves and legs of a foal or a fawn and the developmental mode of its posture, locomotion, and voice. In young harbor seals, the hydrodynamic form of the body takes shape in the same way as do the drive to swim and the correct mode of operation of the paddle-shaped limbs.

Immediately after birth, behavior appropriate to the species begins, in movement as well as in sensory perception. However important in particulars the marking of the definitive behavior by experience with the environment may be, this marking is still limited in advance by the narrowness of the prescribed environment. That is why the life history of animals is so unvarying; a writer seeking to describe animals in a literary way must either completely anthropomorphize them or must interweave their lives so closely with human lives that individual traits stand out and a kind of biography can be read into the description of the animal's life. Research into literary attempts to construct biographies for animals might be very informative on the subject of the great differences between the lives of humans and animals. By no means should individuality in animals be underestimated, nor should the possibility of their having tradition, which adds nuance to instinctual behavior; but the consequences of these eventualities, insofar as life histories are concerned, are considerably different than for humans.

We have taken into consideration here only the higher mammals as representatives of animal life, for in a comparison between human and animal, only these highest types of animal existence have a bearing on the discussion. Now let us compare the developmental mode of humans with the most important features already mentioned in the ontogeny of higher mammals. The most significant contrast is that, at birth, the human has not yet attained the type of movement, the body posture, or the power of communication typical for its species at maturity. Instead of maturing to that stage within the womb, of becoming a juvenile developed to the highest mammalian form, the little human creature is released from the womb at a much earlier stage and "brought into the world." At first glance, such a course of development resembles that of the primitive mammalian altricial level. But we have already seen that this interpretation stands in need of correction and have termed the newborn human "secondarily altricial" because, based on its developmental level, it really must be assigned to the precocials even though it lacks the precocial's freedom of movement.

If the human were construed as a true mammal, the additional length of time it would have to spend in the womb to attain actual precocial formation corresponds roughly, as we have seen, to the first year of postpartum life. When compared with the animal norm, this period appears in a special light. We call it the "extrauterine spring" and shall now turn our attention to this segment of our life.

Three significant events characterize the first year of human life: the attaining of erect posture, the learning of an actual verbal language, and the entrance into the realm of technical thinking and behaving. In the following material, we have selected for emphasis a few important steps from the history of the origins of these three special events. With this choice, we are considering processes that can be understood with particular clarity; it goes without saying that during the same critical periods of time other important processes are also going on that are not as easy to characterize.

Erect Posture

NO OTHER mammal attains it species-specific posture as humans do, through active striving and not until long after birth. Even though in many animals, the first movements are immature and labored, and in some circumstances concealed by a powerful clinging instinct, as in the great apes, the entire body posture and the motor patterns of such forms are typical of their group and, in general characteristics, are similar to those of their parents. The newborn ape has to learn and practice a richer program of species-typical movements than, for example, the newborn ungulate. But it does not first have to mature in bodily proportions, as a human does, and does not have to learn species-specific erect posture at the end of its first year of life.

In humans, neuromuscular organization is already well developed at birth, but does not begin to assume its extensive, final form until months later. This later development, however, does not occur simply through practice, using a predisposition already inherent in the structure, but through special acts of striving, learning, and imitation peculiar to the organism, while the body goes on forming through a very conspicuous, differential growth of its parts. Even the actual bodily structure attains the mark of its species with the help of this striving. The vertebral column, which in the newborn is almost straight, does not attain the characteristic curvature of a resilient support structure for a perpendicular body until late; correspondingly late and with significant shaping processes, the pelvis, too, assumes its typical position. Almost three full years of the child's life pass before the pelvis and the vertebral column approach the form of the mature figure. We give here only a few steps out of the entire process of attaining erect posture, with the average times when they occur:

SECOND TO THIRD MONTH: mastery of holding the head erect;
FIFTH TO SIXTH MONTH: striving to sit up, and succeeding;
SIXTH TO EIGHTH MONTH: standing up, the whole body erect, with the help of adults and supporting self on objects;
ELEVENTH TO TWELFTH MONTH: standing alone for the first

time and undertaking the first independent steps; subsequent rapid learning to stand and walk alone;

ELEVENTH TO THIRTEENTH MONTH: learning to get up from the ventral position.

We are not so concerned with walking upright, as is usually the case, but with standing, with erect posture, for this is what is special and human. Walking in this position is the relatively simple function of a very primitive neuromuscular organization common to all quadrupeds—alternate movements of the limbs. This type of movement is strongly built into the genetic makeup of all quadrupeds: Thus, we observe in immature birds first the alternating movements of the little wings, a primitive and useless type of movement, which only later is succeeded by the simultaneous wing beats suitable for flight, and in many songbirds even by a simultaneous hopping motion of the legs. Stirnimann's observations (1940; see note 4) of spontaneous, finished mechanisms of movement already present in the newborn indicate that the types of movement common to all quadrupeds also appear in man. Stirnimann ascertained that, on the first day of life, 29 percent of newborns, when laid on their stomachs, make crawling movements, and 16 percent, if they are supported in an erect position, make walking movements. If the same group of infants is observed again after about two weeks, at the end of what is called the neonatal period, about 59 percent of them exhibit the crawling movements, and 58 percent, when held erect, the step movements. Throughout the entire period, no opportunity for any kind of exercise had been offered; therefore, these types of movements mature on their own. That they are very primitive follows from the fact that, according to Stirnimann's observations, the movements disappear between the third and the fifth month. Not until nine to ten months do babies once more crawl or walk when placed in those same positions. However, according to Variot, that is the time when 38 percent of children experience the onset of what is called "prelocomotion," which precedes actual standing and taking steps. This stage begins at about seven to eight months and is clearly related to new motor impulses, which, as Variot noted, are occasioned by the advancing development of the brain and mental functions.[15]

The true significance of the slow acquisition of completely erect posture and the necessary supporting physical structures is still difficult to understand. We must still be satisfied with pointing out that the formation of one of the most characteristic marks of the human being is postponed until a time when major mental developmental processes, the formation of our world experience, are also taking place. Perhaps in what follows we will succeed in making apparent, at least in revealing outline, a deeper context underlying this temporal correlation.

In the distinctive process of learning and fixating species-specific posture, the circumstance that the legs of infants are very short, making attempts to stand that much easier, is of considerable assistance. As early as the fifth prenatal month, growth of the legs has already fallen behind that of the arms, although the ultimate size relationship would lead us to expect the opposite. If we see in this early retardation of leg growth a process that is coordinated with the special way in which the human acquires erect posture, we also understand the other striking fact: intensive growth of our legs does not set in immediately after birth, after the conceivably restrictive environment of the uterus has been left behind, but only very gradually after the sixth month of the extrauterine spring. Not until the beginning of attempts to stand, therefore, and, more particularly, after the acquisition of erect posture, do the legs begin to grow more quickly. Such an interpretation of the early lagging behind of leg growth before birth seems to me, in our limited knowledge, to correspond better to the observed evidence than that other interpretation which claims to see in such deferred proportionality the lasting effects of completely unknown ancestral states.

One could hit upon the thought that learning to stand takes

place during the extrauterine period because such a complex behavior could not be developed within the narrow confines of the womb. However, a glance at the development of birds shows that such an interpretation is untenable. Precocial bird babies—those of chickens, ducks, and plovers, for example—form within the egg, despite the close quarters, all the neuromuscular systems they will need for standing on the first day of their independent lives. Therefore, to use quails as an example, such a complex capability can be readied in sixteen to seventeen days without the slightest possibility for practice. Furthermore, the foundations for flying arise in birds while they are still in the nest, without any kind of preliminary practice: swallows and swifts, in spite of the completely different way of life in the nest, bring to their first flight an astonishing mastery of gliding, a movement they have never learned or practiced. We should probably assume from these and analogous facts that in the human, too, walking and erect posture could form within the womb without preliminary exercise. That this does not happen, however, indicates that this act of our development is intimately associated with other typically human developmental processes and can only be completely understood within such a context.

Language

EVOLUTIONARY THEORY has introduced into biology, a science primarily oriented toward zoology, the assumption that human speech developed from animal sound formation through gradual enrichment and transformation in the meaning of the sounds. Such attempts at a purely verbal derivation casually overlook the difficulties in principle posed to every kind of research in natural science by the problem of origin. This schematic mode of thinking explains the ontogenetic events in learning speech as a repetition of the assumed evolutionary process, as a gradual transition from animal sounds to human speech. Corresponding to this basic concept, physiology textbooks go on to treat "speech" quite simply as if it were identical with the means of producing sound. So it is that right from the beginning of a biologist's education, a disastrous confusion is established that helps to conceal one of the great problems of anthropology.[16]

For that reason, attention must be drawn emphatically to the fact that in humans, both verbal and gestural language is based on the communications principle called "signs," something completely different from all animal sounds. All attempts at derivation, therefore, lead from animal sounds only to the human sounds that correspond to them—to the cry, or call, to take the most conspicuous as an example. Based on the pattern of all deductive strategies, all theories of descent are compelled to explain the word as a bearer of meaning in an entirely verbal way, as a "very gradual" formation. Every attempt of that kind remains, however, a purely intellectual invention and furthermore, in order to work out, has to be based on a stage of human development that, with regard to the mind, is already "human."

"We understand speech as the function whereby, with the help of organized structures of sound and sign appearing in various meaningful combinations, we are able to describe our perceptions, judgments, wishes, and so on, with the intention of sharing them with others, of mutual understanding" (Revesz, 1946; see note 16). However impressive and "meaningful" many animal sounds may seem to us, they are, like our cries, always only expressions of inner states, and not language in the sense just described. Only when the misleading equating of sound production with speech, of animal noise with word, is overcome in biology will the special nature of what actually takes place in human children as they learn our verbal language be seen correctly.

First, we shall once again describe the steps of this event as they have already been inferred from simple observation:

Preceding any acquisition of language is the human capability —genetically predetermined and always at hand later as a possible reaction—of crying, growling, squealing, or clicking, which is to say, of producing general expressions of inner states. We shall not discuss here the difficult problems of laughing and crying, for language learning can be observed independently of the study of these phenomena; nevertheless, it is clear that these problems must be topics in any complete biological discussion of human behavior.[17]

In the third to fourth months, the child begins the manifold attempts to make the movements with which, especially in the fifth and sixth months, it produces sounds. This exercising leads to babbling, to actual monologues of babbling, with which the tiny creature produces a vertiable arsenal of sound units, many of which it will no longer use when it learns its native language and will have to master again when it learns a foreign language. This phase, with its wealth of elements, contains the possibility for learning any human language whatsoever.

At nine to ten months, however, imitation of words from the social environment begins. This imitative speech can remain very inadequate for a considerable period of time. Words are mimicked that relate to facts, and, at the outset, they can also represent very varied, complex situations. The words stand for wishes and strivings just as much as for assertions; they represent virtually entire sentences. Much richer than the words spoken are the mental processes, the multiple aspects of which we infer based on the details of the situation. At this early stage in life, the concealed development of the mind is already much richer than what can be expressed in words; the limits of our means of expression at this time are already an indication of one of the most important limitations of all our social interaction, of our entire human existence. The actual acquisition of speech—which follows the production of sound toward the end of the early postnatal period—is the imitative adoption of a complete, preexisting social apparatus, a process that is deeply intertwined with the child's life as a social being and continues for a long time with great intensity.

The incomprehensible event of human language learning during childhood development becomes impressive when a child is compared with a young chimpanzee. In Moscow, N. Kohts (1937; see note 3) identified twenty-three different sounds with expressive possibilities in a young chimpanzee more than one year old. She found that her own son, Roody, could produce all of these sounds when he was just seven months old. But in the eighth month, Roody also imitated a human word, and at fifteen months he used words to name objects. The chimpanzee, on the contrary, never made the slightest attempt to imitate any sound recurring regularly in the surroundings.

No one underestimates the wealth of animalian means of communication, as revealed by behavioral research, or the multiple possibilities for relationship that go along with different functions in the social group. Nowhere, however, do we find the possibility for using a word as a "sign," freely disposable, independent of a particular situation. What we know today of how higher animals effect social relationships by means of gestures or sounds, what has been found out about the communication system of bees, for example, leaves us with a much higher opinion of animals than ever before. But it is precisely because of such research that the uniqueness of our language seems grander and more mysterious.

Insightful Behavior

JUST AS striking as the preparation for standing and the mimicking of the first verbal structures is the behavioral transition from simple unreflective imitation, which appears early, along with instinctual behavior, to actual insightful behavior. That even imitation is coupled with momentary acts of insight and understanding is just as clear during this early period of childhood development as it is in chimpanzees. For this reason, the "aha!" experience is presented by psychologists as the one of the most interesting borderline accomplishments of chimpanzees. However, the decisive factor in human children is the ultimate surpassing of this stage at about the ninth or tenth month of the first year, the attainment of a level at which insight, the understanding of meaningful contexts, becomes a typical element of our behavior. This insightful behavior begins with the grasping of the concept of tool-using, with technical intelligence. It begins with the transference of the solution of one problem to an analogous, yet very different situation, whereby the child progresses from undifferentiated behavior—from subjective to objective understanding. In this recognition of analogous situations in spite of very different background factors, a very special, significant expression of humanness is manifest.

Oneness of Developmental Events

THE THREE phenomena of the first year of life that must be isolated in our presentation do not appear separated in this way in the child's development. Conspicuous simultaneity in the appearance of individual steps bearing related features reminds us that it cannot be a matter of chance concurrence of independent courses. Important commonalities show up in these very different processes, which must lead to the conclusion that one, deeply concealed cause is at work.

A first concurrence appears in the role of the first stages of neuromuscular development, which precedes the acquisition of erect posture, and of speech and behavior as well. We see the child actively striving toward ever newer postures and movements, in the area of the torso and limbs as well as in the larynx, which is hidden from view, and in the area of the musculature of the tongue. All these efforts at movement lead not only to the satisfaction of an urge to move, but at the same time to an intensive knowledge of the child's own body, to control over the movements of the arms and legs and the so very important little fingers, and also, and in the same way, to power over the movements of the organs of sound. And so, as the small infant in the course of this incessant activity gradually recognizes his body as "his" and understands the wealth of possibilities at his disposal, he also experiences through his own acts the possibility of hearing sounds he himself produces, sounds that are also "his." Here, we only hint at the significance of this important early period of experimentation with movement for all later experience and the demands it will place upon capabilities.

All these processes are signs that concealed events are taking place within the child's central nervous system; but—and this is critical—they are not just signs, but rather themselves elements of the developmental process in that they continually produce new relationships and create with every act completely new points of departure that could not have existed shortly before, leaving behind a new context for all further events.

We cannot present clearly enough the contrast between the development in the womb, which entails processes that are so alike for all individuals of a species, and the special developmental conditions that create such a wealth of possibilities for the springtime of the human child—and all during a period of time when such significant events are taking place in the shaping of our bodies. In humans, maturation processes, which did indeed begin within the mother's body, go through their most important phases in combination with the experiences offered by a much richer environment with many sources of stimulation to the organism capable of development. Thus, in humans, courses of events ordained by natural law take place during the first year of life not in the all-purpose environment of the womb but under unique circumstances; each phase of postpartum life intensifies this uniqueness by increasing the possibilities for divergent, individual situations. And so it is that already during its first year of life, the human child is subject to the laws of "history," at a time when the human as a true mammal would still have to be developing within the darkness of the womb, in conditions governed exclusively by natural law. Even during this extrauterine springtime, not only do "processes" of the most general kind take place, but also countless "events" that are unique—and often fateful, although in the context in which they appear we are not completely able to assess them as such.

Along with the spontaneous urge to try out new movements, another trait common to the three main human developmental processes of the first year is the great significance of imitation of the modes of behavior presented by the social environment. Whether one traces the origin of erect posture, of verbal language, or of the development of insightful behavior, one always encounters the role played by imitation. Early on, it begins, in a manner of speaking, to seize upon the experimental movements and more or less to orient them. Moreover, one becomes aware, from the very beginning, of the extent to which these human qualities of posture, language, and behavior are phenomena with a social stamp, and it becomes obvious how much circumstances of social contact contribute to their formation right from the outset. The continuous, permanent interaction between help and encouragement from the surroundings and the child's own creative activity and urge to imitate mark the developmental course; all these factors contribute equally to the development of physical characteristics and to the way of life.

Ungulates and apes, seals and whales—all mature in seclusion, in the mother's body. Perhaps this assertion strikes us as doubtful, given what we know of the juvenile period of higher animals, of the period of experimentation and learning within the group, of the impressive early phase of an ape's life, and of the life of an anthropoid ape in particular. That is why we must consider again the singular nature of our own birth state, the high multiplier factor of our brain, and the different proportions of limbs and torso in order to understand the uniqueness of our early situation. For a moment, we must focus intently on the situation we find when we study mammals other than humans to establish the developmental norms of this animal group; we must try to imagine the developing human spending the important maturation period of its first year of life in the dark, moist, uniform warmth of its mother's womb. Only then, by contrasting that vision with the reality of human development that is before us, will we understand the completely special, separate nature of our mode of development. Then, as we reflect on these things, the unusual, intimate relationship that exists between the special nature of human behavior and the remarkable, atypical development of our children will become apparent step by step. It will gradually become clear that world-open behavior of the mature form is directly related to early contact with the richness of the world, an opportunity available only to the human!

Perhaps another look at the nature of this correlation is still necessary. We are not saying that early contact with the world is the cause of typical human behavior; what we assert is a relatedness of phenomena, a correlation between the existential form of an organism and its ontogeny. The singular nature of human development gains meaningful interpretation when set within the context of the entire human existential form.

Form and Behavior as One

FORM AND behavior—aspects of the human being that are inextricably bound together, and separated only by us—diverge decisively from the familiar developmental mode of animals. In humans, both aspects mature and develop not just in the protected environment of the mother's body, and not just along lines laid down by inherited developmental laws laws that shape the being to conform to its future environment. It is allotted to the human that it experience critical formative phases of behavior and bodily growth outside the womb, with close interaction of mental and physical events. The human is truly "abandoned" much earlier than any other mammal, ousted into the fullness of its intended environment. And as it, in a manner of speaking, grows into its environment, the special nature of the upright individual and of human world experience comes into being.

Human otogeny, which differs so considerably from the mammalian norm, is connected to the unusual nature of our behavior emphasized earlier; it corresponds to the situation of a creature open to the world, to the circumstance that our social world is not given to us genetically, but must in every single human being be shaped anew from inherited structures and contact with reality. The peculiar nature of the acquisition of speech, the imitative appropriation of an entire preexisting, richly structured social instrument—one composed of speech, of gestures, yes, but also of technology and of opinions—is in the highest degree emblematic of the facts of our case. Our mental structures do not mature through self-differentiation to become finished behavior patterns, capable only of the slightest subtleties, as we know maturation to occur in animals. In humans, only when these structures come into contact with the rich content of the surroundings do they unfold to become the form, shaped by its time, that will be characteristic for each individual. This special, different kind of development is guaranteed by the fact that the human is indeed born in a very advanced state with regard to both structure and mind, and yet its forms of behavior are far from mature; for them to develop, the possibility of contact with the surroundings, of being exposed to the richness of the world, and of experiencing social interaction is created.

Let it be emphasized once again that the formation of the most significant structural and mental peculiarities takes place during the course of one and the same postpartum developmental step, in striking simultaneity and with the innermost correlation of events. And yet, how often is the body regarded only as the material basis upon which the further potential of real human existence rests; this body is taken as a vessel within which higher human qualities will later develop. Even evolutionary theory has given strong support to the concept that first, the human form developed from preliminary animal, apelike states, and that its final flowering was the human mind, which ultimately bore the fruit of culture.

However, our developmental process shows a connection between bodily structure and behavior that is much more intimate than the relationship between container and contents. In our process of becoming, the world experience characteristic of us, as well as the final form, which is ours alone, develop as an indivisible unit, in continuous, exceedingly close interaction. We use the neutral word "develop" *(entstehen),* for if we said "form themselves" *(sich formen),* the expression commonly used, we would be saying more than we actually observe. Every organic form encompasses the mystery of synergy, of reciprocal influences of the outer and inner worlds that are difficult to comprehend. How deeply will scientific research be able to penetrate these events? Whatever the answer to that question may be, we want to try to see at least the unusual nature of our development, which provides important, basic evidence for this reciprocity. Perhaps such observations will also set the stage for the understanding of relationships that are more difficult to discern but related in kind.

It is widely believed that our ontogeny passes through stages of the organic kingdom, from the unicellular egg to the human being; that the human, after its germ has traversed the lower stages of organization in brief, rapid succession, finally becomes a mammal, then a primate, an ape, and, emerging from the chimpanzee stage, ultimately becomes human—and the timing of the appearance of the last event in this process is set variously by biologists. That these concepts come to us in the guise of scientific truth—often under the great name of biogenetic law—should not delude us into believing that they are the last word!

Certainly, the separating of many of these stages has directed attention to important characters of individual developmental stages; but the exaggerated importance given vague, formulaic similarities in the theory of evolution (such as correlating the one-celled "protozoan" with the egg cell), has sometimes been more of a hindrance than an advancement of knowledge. If, instead of allowing our attention to be led astray by diagrams of stages, we turn our attention to a careful observation of what in our development is particularly ours, an essentially different picture will appear. Then we comprehend this quality of ours that is so special, so human, in the early stages, too. And if it is also concealed at those stages, then, as when man-made things lying beneath a thick blanket of snow reveal their presence only by surface contours, so can these indistinct features induce us to observe more intently, to perfect the means that enable us to interpret just such concealed traits. The same is true for the developmental history of animals, for more sophisticated observation will also considerably broaden our concept of the distinctness of animal groups.

To interpret development from the egg to the mature form by means of a diagram has become dangerous, not because all it requires is a sort of simplified, easily remembered presentation, but because it has been understood to express a law of recapitulation of ancestral stages. The danger in this sort of diagraming was fully recognized as early as 1873 by W. His; he vigorously opposed the diagram, as did L. Rütimeyer,* and also described the unusual nature of very early mammalian developmental stages, supporting his work with careful measurements. Precise studies of proportion, if they are not fettered by an exclusionary concept of evolution, would also show significant individuality for the different human fetal stages. Both studies by Schultz, which have already been mentioned (see note 5), constitute good beginnings in this direction.

Observation of the mental development of infants and small children leads again and again to the fact emphasized in this study

*Wilhelm His (1831–1904), German anatomist and embryologist, author of *Anatomie menschlicher Embryonen,* published in 1880–85. This book is considered the first accurate study of the development of the human embryo.
Ludwig Rütimeyer (1825–95), Swiss zoologist, author of *Die vorweltliche Tierwelt der Schweiz* (Prehistoric fauna of Switzerland).——Tr.

that the nerves and the senses have attained a significant degree of formation by the time of birth. This truth compels attention to the time before birth, so much more difficult of access, which creates these conditions. We must assume that early on, within the mother's body, elements are present that, although hazy, already exhibit the special marks of a human being and that should not be appraised as being simply mammalian stages or, in later phases, as ape stages. Comparative embryology, misleading even in the earliest phases of ontogeny, should be replaced by another model. Let us examine the model of artistic forms to see if it corresponds better to what happens:

A picture begins with a sketch, its first, faint lines on the blank paper already implying the finished whole, unclear as the details may still be. As at each subsequent formative stage important traits of the future work emerge, clear in some places, not so obvious in others, so in the development of a human being are the traits that are crucial to the special nature of that being present right from the beginning; "humanness" is present from the very first. How many developmental steps are already set up to provide just for the human peculiarity of early birth, for the special nature of our later attainment of species-specific posture and speech!

We are trying to present a metaphor. It is just an image, one that can perhaps help to direct our thinking in fruitful research directions. We do not want to forget that even the egg, with its simple, spherical shape, is not just a visible, simple ball-like object, but is filled with the potential for human form. It is on the way to becoming that one, particular human being, and absolutely nothing else. Let us stay a moment with the comparison between the developing organisim and the emergence of a picture from a sketch. In the unfinished drawing, each stage of development still includes the possibilities for all kinds of variations that will be eliminated as the picture nears completion, as one path is followed. In the same way, the developing germ on its way to a final form is unfinished and suggests other possibilities of formation. But that is exactly why, because it suggests these possibilities, that at such stages it is not a fish, a reptile, an ape.

Only by seeing in each of the developmental steps the potential for becoming a human being, an organism with unique, erect posture, with a different, special kind of world-open behavior, and with an innate capacity for participating in a socialized world of culture shaped by language will we gain a more profound understanding of human development.

5

Growth After the First Year

THE LONG duration of our growth period must be particularly striking in any comparison of human and animal development. Whereas the unusual nature of the first year of life has attracted little attention, the longer, subsequent period of growth has been taken notice of and interpreted again and again. Usually, our growth mode has been described as "delayed" in comparison with that of an animal. Correspondingly, retardation and the related concept, fetalization, have also recently become key words for theories of anthropogenesis and all biologically oriented anthropological research. In this description, we want to begin with the verified facts that the comparative survey has yielded. Only secondarily, deduced from the always verifiable evidence, can an interpretation in the sense of an idea about evolution follow. Therefore, we prefer a descriptive term for human growth from the end of the first year of life until the onset of puberty, saying that it is slow and not actually delayed. This second assertion requires a special line of argument.

The Singular Nature of Later Human Growth

FIRST, WE mention a few facts, which, when surveyed together in this way, will bring out the special nature of our mode of growth after the first year. In doing so, we limit ourselves to the large, more highly organized mammals, which are the only creatures whose level of development can be compared with that of the human.

Even the casual observer is struck by the impression that for the large mammals, growth in mass is usually quite rapid. For example, a red deer is full grown at three years of age, and a lion at six to seven years (sometimes even earlier), and, for both animals, growth during the last of those years is very slight. Most of the increase in size has already taken place between one and two years, after which time the animal is usually capable of reproduction even though growth has not stopped completely. Growth patterns are similar for all the larger mammals; even the hugest of them attain their size much earlier than is generally thought. A young blue whale, the largest of all mammals, is seven meters long at birth, doubles its length in just seven months, and by two years, has most of its growth behind it. At that time, a male blue whale is 22.6 meters long, and a female, 23.7 meters, on the average. After this enormous increase in mass, growth proceeds more slowly, adding about two to three meters, and the ultimate size is reached, based on the lowest assessment, at between five and six years. Again, the critical increase in mass to the mature form takes place during a period of enormous growth immediately after birth. Even for elephants, which grow much more slowly than whales, the only reliable figures we have point to the main period of growth lasting only until the fourteenth and fifteenth years, even though the intense early growth period lasts longer than it does in whales. All figures for older elephants are within the normal distributional range for the finished form.

Particularly important for us are the situations found in higher primates (see note 3). Today, we have more exact knowledge of postfetal growth in macaques: according to Schultz, the growth period lasts six years and five months. The much bigger gorilla grows just as rapidly; female gorillas reach their final size in just six to seven years. The male is full grown at about ten years. Chimpanzees stop growing at ten to eleven years, and the time is about the same for orangutans. But in these species, too, the major growth is finished much earlier; chimpanzees reach a size that is within the adult range when they are eight years old, and orangutans, when they are between nine and ten years old. In contrast, the human growth period lasts nineteen years all told, if we take only the postfetal time into consideration; more recent figures compiled here at home have the period lasting to the twenty-fifth year.

However, simply stating the length of the growth period gives an inadequate picture of the remarkable nature of this period of human life. Only when the growth period is broken down into segments does our unusual situation show up clearly for what it is. All mammals (other than humans) grow very rapidly right from the start of their independent lives, and have the major part of their growth behind them by the time they become sexually mature. Any growth still to come is slow and slight. In humans, on the contrary, growth processes experience a marked increase in intensity at the very moment of sexual maturation, and it is during this late phase that a significant part of the total growth takes place.

Whether there is anything in the great apes that is really comparable to our growth at puberty cannot be ascertained with any certainty based on available evidence, although figures compiled by Spence and Yerkes (see note 3) for male chimpanzees show that a slight increase is likely. Evidence upon which to base a final assessment of ape growth is still very scanty. Even the careful accounts made by Spence and Yerkes are insufficient, for in this study, too, the problem of the median weight of adult chimpanzees has not been cleared up. In the different studies so far available, a range of figures has been given for the median weight:

COUPIN	35 kg
SPENCE AND YERKES	39 kg, female; 46 kg, male
RODE	55–65 kg, female; 55–75 kg, male
SCHULTZ	ca. 75 kg
BRANDES	ca. 75 kg

Depending on which of these figures is chosen, a growth curve would naturally indicate very different times for the attainment of full growth. Of prime importance for our problem is the following: 1) growth in orangutans and chimpanzees, the slowest among primates, and that of elephants, the slowest of all of the higher mammals, approaches its termination at the point where, for humans, the sharp increase in mass at puberty is just beginning; 2) the high point of our growth, which occurs at the fourteenth to fifteenth, or fifteenth to seventeenth years, takes place in apes well after growth has ceased.

In connection with the increase in size during puberty, we must also look at the time when sexual maturity occurs. In female macaques it appears as early as the third or fourth year and, in males, sometime during the fifth to sixth years. Female gorillas mature in the sixth year, males in about the tenth. Chimpanzees become sexually mature in the tenth or eleventh year, and, once again, the male is later. Late maturity is characteristic of all higher primates. These values can only be properly assessed, however, when they are compared with values for other large, higher mammals. Deer are already mature at one and one-half years, cattle at one and one-half to two years. The elk, like the donkey, but also the massive hippopotamus, is ready for reproduction between two and three years, and the horse, between three and four years. Careful observations in zoos show us that female elephants become fertile at seven to eight years, and often not until the ninth or the tenth year, whereas the male can procreate at eight or nine years, often not until ten years. As recently as 1917, the information that elephants required twenty to thirty years to reach maturity could still be found in biological works. Here, even in our disenchanted age, hoary tradition is still at work, for it is primarily the giants among animals who remain repositories of ancient beliefs about animal life! Even Brehm's *Tierleben** says that the huge baleen

**Tierleben,* by Alfred Edmund Brehm, German zoologist, 1829–84. This six-volume classic on the lives of animals was first published in the years 1864–69; a fourth edition in thirteen volumes appeared 1911–18, and popular editions continue to appear.——Trans.

whale hardly seems capable of reproduction before its twentieth year. Studies conducted over many years, which have served as the basis for the modern whaling industry, show not only that the large whales grow rapidly, but that they are able to reproduce at between four and six years.

Only in the light of these figures for higher mammals can the exceptional situation for humans be recognized. When our late sexual maturation fitted into place between the figures for the great apes and the extraordinary times given for whales and elephants, it had to appear simply as one of the late maturing types of mammalian development, whereas it turns out today to be an exception, or at least the absolute extreme in the series of figures given for mammals. We give here as the norm the situation as it was in the Western world before the striking phenomenon of "acceleration," a norm still more or less valid for peoples living in less technological societies. The median age for the onset of maturity for the white race (on which our information is always based, unless otherwise stated) is as follows: for males, thirteen to fifteen years (the onset is difficult to establish precisely through a temporal event); in females, the fifteenth year (beginning of menstruation) may serve as an average time for the admittedly very variable maturation point.

We are able to identify in general and also more specifically many factors that contribute to the attainment of puberty and to the overall distinctive nature of human growth. For instance, we understand the growth-enhancing function of the thymus during the period before maturation; we suspect that during the same period of time there is an inhibiting influence exerted by the structure known as the lymphoepithelial pharynx—that is, a ring of lymph glands in the throat, of which the tonsils are a part—and by the palatine tonsils in particular. (The palatine tonsils have their maximum development at six years, the pharyngeal tonsils at ten years, and both organs later degenerate.) With the onset of sexual maturity, another antagonism becomes quite pronounced: the growth-enhancing effect of the thyroid gland begins, stimulated by special hormones of the anterior lobe of the hypophysis—the pituitary. Operating counter to those hormones are the inhibiting effects of the sex glands, the female hormones seeming to take effect more forcibly than the male components. Very difficult to understand is the inhibiting effect assumed by many biologists to be exercised on the gonads during the early years by the hormones of the pineal gland (epiphysis cerebri). The influence of the adrenal cortex, the details of which are as yet insufficiently known, is one of enhancement.

Nevertheless, the role played by all these substances provides no key to a deeper understanding of the events that are especially typical for us humans, for all the substances produced by the ductless glands are also formed in the bodies of animals, where they also influence growth. (Indeed, we know many of the ways in which they work in our bodies primarily, often exclusively, through research on animals.) It is true that insight into the workings of such substances explains general processes of how growth is directed and satisfies the biologist who is primarily looking for general laws. We have learned something about a system of releasers and control factors that occurs in all higher vertebrates. However, all this knowledge tells us nothing about the peculiarity of the growth processes in our own bodies. We establish only that this general system of releasers and controls directs the tissues and organs according to a plan that is typical for each species, and that, therefore, the species-specific temporal succession of effects is genetically fixed. But this very temporal deployment of growth-enhancing and -inhibiting effects is an essential part of the singularity of our growth processes; nevertheless, for the time being we can only describe hormonal influences, and cannot explain them through causal analysis. To what extent we will eventually be able to do so is for the future to decide.

The Growth of the Brain

IN CONDUCTING research into anything as complex as an organism, biology does not limit itself to causal analysis but also makes use of the possibilities offered by biological comparison. Comparison, used at first to reveal structure, has already led to an important observation just in the consideration of fetal development. It appears that human development of mass before birth, which results in our newborns being much larger than fetal anthropoids, can be meaningfully placed in context in only one way. At birth, the brain mass of a human exceeds that of a newborn anthropoid by approximately the same degree that the mass of the adult human brain exceeds that of the adult anthropoid brain. Furthermore, it turns out that the rest of the body mass of our newborns, considerably larger than what we see in the great apes, can only be understood as an approximate adaptation of bodily proportions to the considerably larger brain mass.

The conspicuous preponderance of brain development observed in the earliest stages of all vertebrate life is in higher mammals sustained long into the developmental process and appears very clearly in the formation of humans. A causal relationship has been established between the considerable size of the adrenal glands, particularly their cortices, during fetal life and the advancement of brain growth. We observe that in the third fetal month, the adrenal body is the largest organ of the entire abdominal cavity, and the corollary has also been observed: that in more than half of the instances in which the adrenal body is lacking or severely undeveloped, there is conspicuous inhibition of brain formation. It seems as if the adrenal cortex must be one of the structures that stimulates the growth of the brain during its early stages.

We learn from Variot's observations on the beginnings of standing and walking in relation to age and general bodily development in babies that, during these early developmental process, anabolism in the brain is favored. Variot shows that the point at which standing and walking begin is definitely not related directly to the length or weight of the body as a whole: heavier, larger children do not stand sooner, and slightly built children do not stand later than the average for babies. Rather, the onset of this event is dependent on absolute age. Since it is a question of a process directed primarily by the organization of the central nervous system, this temporal constant shows that even in very different types of bodily development, the central nervous system develops at the most even rate of all the organs.

Manouvrier's* observations demonstrate the same thing. He calls attention to the fact that, early in an infant's life the brain can continue to grow and develop normally when the rest of the body is, for whatever reason, losing weight, and that the brain weight of malnourished children is often normal, or at least that the brain is never proportionately underweight.

Now, we should like to point out another phenomenon that is perhaps explained in relation to the enhanced growth of the brain: the likelihood of a closer relationship between brain formation and second dentition. The milk dentition of anthropoid apes appears soon after birth and is complete shortly after the end of the first year. Since 1966, we have had more precise information on growing gorillas from Lang.[18] The milk dentition first appears at seven to eight weeks and is complete at twelve and one-half to fifteen months. Humans do not cut their first tooth until the sixth to the ninth months and finish cutting teeth at twenty to twenty-four months—around the end of the second year; it takes a full year and a half for all the teeth to come in. In macaques, the first definitive cheek tooth appears during the second year; in gorillas, not until the end of the second year; in chimpanzees, during the third year; in orangutans, during the third to fourth years. But in

*Manouvrier, Lóonce, 1850–1927. French medical doctor and anthropologist.—
—Trans.

humans, four whole years pass between the completion of the milk dentition and the beginning of the second dentition. These dates appear to be in some way related to the growth of the brain; the dates are not as precise as we would wish due to the difficulty of measuring the processes, yet it seems to me that the possible significance of these dates is important and that they deserve greater consideration.

For example, if data—scanty as they may be—on brain growth in apes and great apes are compared, the striking coincidence appears that in all forms, the brain has attained at least 90 percent of its mass by the time permanent dentition begins. Comparison of braincase capacities supports this statement, which is based on weight data. For humans, the statement is based on the careful measurements of Roessle and Roulet (1932; see note 12), which show that after the fifth year, the weight of our brain falls within the range of that for the adult organ. If we compare this with other data on brain growth in humans, the cessation of increase appears somewhat later, during the seventh year, for example. We should, therefore, probably venture the supposition that in humans, too, about 90 percent of the volume of the brain has been formed by the onset of permanent second dentition. The figures A. Schultz used to summarize the development of braincase capacity support such an assumption, even though they do not relate exactly to the stage in question. Table 5.1 shows not only that, toward the end of childhood, the difference in relative size (as compared to the mature stage) is abolished; the figures in the second column also show how rapidly the brain develops after birth. On the average, 83.55 percent of the definitive brain weight has taken place by the end of childhood, and 96.1 percent of the brain mass is present by late adolescence.

Variot (1926; see note 15) also points out a particularly close relationship between brain development and dentition. He shows that the beginning of standing alone in infants is not directly related to overall growth (we have already raised this point). On the contrary, his observation shows that 50 percent of the children he studied took their first step at the time when six to eight teeth had been cut—and the timing had nothing to do with the rest of their bodily development.

Only after the period of significant brain growth and in connection with the appearance of the permanent dentition does the jaw begin to grow; in the great apes, this leads to the rapid formation of the animalian snout and in gorillas, goes so far that the outer contour of the jaw at the milk dentition stage becomes completely enclosed by the inner arc of the adult jaw. We see in humans, too, the same kinds of temporal relationships in growth phenomena as are apparent in anthropoids, and we should probably infer a closer linkage between these processes. But further development also presents us with something uniquely human—the striking deviation in later growth in the jaw area. In humans, the zones of the jaw extending beyond the milk teeth continue to grow, providing

TABLE 5.1.

	New Born	*Late Childhood*	*Late Adolescence*
macaque	40.9	79.8	91.2
gibbon	62.7	84.3	95.5
orangutan	40.4	90.4	96.8
chimpanzee	45.3	83.7	98.8
gorilla	59.4	81.4	97.9
human	23.3	81.7	96.7
Average	—	83.55	96.1
ape	49.74	} Average	
human	23.3		

NOTE: The figures in the first column illustrate once more the great difference between the birth state of the brain of all apes and that of man. As ever, the information for newborns is relatively scanty, but the contrast is nonetheless clear.

Figure 5.1. Gorilla. No one can escape the effect produced by seeing the mother ape with her baby. Since the earliest times, the certainty that we are related has colored our relationship to this puzzling creature. In all higher primates, even in gibbons, guenons, baboons, and macaques, the bonds between mother and baby are just as impressive as the one glimpsed here, in the picture of the mother gorilla, Achilla, and her baby, Jambo, in the Basel Zoo. This close bond is very strong in all groups of apes, even in the wild. In macaques, it lasts about a year and in baboons, somewhat longer. In chimpanzees, the bond ceases to exist at about the time of weaning, so it lasts until the end of the third year; in gorillas, the bond lasts just as long, but nursing ends after only one year, as G. Schaller reports. In gorillas, later childhood, the juvenile period, extends a little past the end of the fifth year for males, and until the sixth year for females. In chimpanzees, this period is longer by about one year for both sexes. Photograph: P. Steinemann, Zoological Garden, Basel.

space toward the rear for the rest of the cheek teeth, and in the forward lower part, the prominent chin so characteristic of our facial structure is a consequence of increased growth. In the zone where the teeth are, however, the alveolar section of the jaw stays the same size, so that its final contour coincides with the one in existence at the time of the milk dentition instead of surrounding it, as happens in the apes. Using many measurements, Bolk (1924; see note 6) described with precision this growth peculiarity of ours, previously observed by John Hunter*. We do not want to go into the theories that attempt to explain why that section of the jaw does not continue to grow by relating it primarily to erect posture and to the associated special distribution of the weight of the head over the vertebral column. We note only two important facts:

1. For the higher primates as for the human, the most significant brain growth after birth takes place at the time when only the milk teeth are present; by the time the first permanent cheek teeth appear, growth of the brain has essentially come to a halt.

2. In apes, significant growth of the jaw accompanied by formation of the snout begins in the period following the most intense brain growth. In humans, there is no growth in the alveolar section of the jaw.

So far, we have only spoken of the fact that, right from the start, metabolic processes give preference to the brain, assuring it of very even, regular development. It means a further step into a realm where access is already difficult if we conclude, based on that statement, that this favored system is itself a dominant, guiding organ significantly affecting the development of the body. That this guidance role can only be a relative one, should be inferred from the previous description. We have again and again called

*Hunter, John, 1728–1793. Scottish surgeon, founder of pathological anatomy in England. An early advocate of investigation and experimentation, he carried out many important studies in comparative morphology.——Trans.

attention to the initial oneness of the entire protoplasmic body, which, through its invisible morphology, subdivides into organs, assigns roles to these organs, but remains itself, in a way we do not yet understand, the highest authority. Only in this sense, then, do we go on to speak of the guidance role of the brain, which is altogether an instance of the leader that is led. We see this even in the first year of life, in the strength of the instinctual drives, of the striving to stand erect, to speak, and to experience the world. Variot, too, emphasized in his study on prelocomotion that the entire development of mental organization, with its sensations, wishes, and other impulses, creates an increased need for movement that considerably precedes the attainment of actual human modes of movement.

The period just characterized by the interrelationship of brain, tooth, and jaw growth, which lasts until the sixth year, also shows an inner unity; it proves to be a special kind of life stage, an observation that should assure special consideration for the physical events just pointed out. Not by accident is the sixth year the end of babyhood and the beginning of formal education. That such a decisive social act should coincide with the dividing line between these two segments of life reflects practical experience with mental development. The ordinary event of going to school for the first time is the first step in a multifaceted course of events that has been called, from an overall point of view, the first metamorphosis. It lasts about one to one and one-half years and introduces an important developmental stage.[19] When, today, very dubious notions of progress and mechanical learning techniques lead to thoughts of acceleration and simplification of all formal education and attempt to rationalize the preschool age, the biologist can only warn that great care must be taken with respect to such tendencies.

Now, after these references to somatic events, it is time to pursue mental development in its relation to the processes already considered.

Mental Development

DURING THE first year, the great act of standing upright has taken place, and the most characteristic units of language and behavior have been attained; in the period that follows, which we shall now consider, there are important developments in mental activity.

Actual talking—the adoption of adult verbal language—begins at the end of the first year. Often, the new capability is really not put to use until the middle of the second year, and sometimes very hesitantly. Quite often, after the child has learned its first few words, a period of quiescence is observed, which can last from the twelfth to the sixteenth or seventeenth month, the cause of which is unexplained. This lack of progress has been ascribed to natural laziness; perhaps a better explanation is that it reminds us that many inner maturation processes must take place in the brain for the adoption of linquistic forms to be complete. After this period of hesitation, however, vocabulary increases rapidly. The child enters deeper into the stage in which words are understood to be bearers of meaning, where their special connection with things is perceived, and, consequently, all things are "called" something. That is an enormous step, and we shall perhaps be more conscious of its magnitude if we bear in mind how many adults continue to be satisfied with the deep gratification that comes from knowing a name, with naming things already in existence, and, if we think about it, how strong even in more highly developed individuals the effect inherent in the simple naming of a thing remains.

The development of speaking through further adoption of the linguistic heritage of the social group requires much time, which is not surprising considering the wealth of environmental situations available to us humans. In this connection, let us not forget that even languages spoken by so-called primitive peoples are never simple. During the third and fourth years, the child, reaching the stage at which it notices the more subtle variants of word usage, starts to imitate those usages, too. At that point, speech becomes normal and "correct" in the adult sense. But the cost of this progress is the relinquishing of qualities much more secret and creative in favor of something already existing in the environment and, therefore, more powerful. We should not assess this loss lightly, for even in the creation of new knowledge and perceptions by scientists and poets, being firmly tied to a language proves to be a considerable obstacle, and the breaking open of customary verbal structures, the creation of new verbal vessels to hold new meanings, is one of the most difficult tasks of intellectual creativity. Developmental research has shown that after the fourth year, there is a considerable decrease in linguistic perceptive capability.

During these years of improvement in linguistic skill, the child increasingly expresses its need to depict or portray experience. It begins to draw, an activity that has attracted much attention since the time, around 1900, when there was somewhat premature talk of a century of the child, and, more particularly, in recent decades. The mood of the time, one of optimism based on a biologically grounded faith in continuous advancement, foresaw a paradise unfolding in the child's world of play where adults, too, could live a playful existence!

But the representation by drawing that children do is always important evidence of spontaneous inner activity, whether the efforts are the object of loving attention or remain unnoticed in the shadow of more serious parental concerns. Drawing is all the more significant in that it proceeds directly from an innermost urge to create and cannot be conceived of as imitation triggered by the social matrix as much as speech is. The intense individuality of the babbling phase turns up again in drawing. These efforts at graphic representation are also important for their application of the skills in tool using learned during the first year, for their use of the hand, and for their expression of objective experience, of which the depiction of objects is such an essential characteristic.

Around the beginning of the second year, we observe the first signs that images are recognized. Now, the representation is seen as a likeness, as something in its own right, as being different from that which is represented, whereas before, image and thing, insofar as they were recognized as being similar, were confused. This recognition of something as a copy—what a step of mental development!

At about a year and a half, attempts to scribble slowly begin to appear; in many ways, these efforts resemble babbling. Here too, there is much practicing, much neuromuscular testing of situations, of possibilities for arm and hand movement. Facility increases, the movements become more harmonious and simplified, the guidance of the hand shifts from the elbow to the wrist. We must recall from time to time how much this single advance involves concealed events. In scribbling, as in babbling, we observe the production of spontaneous structures the meaning of which we do not understand, but behind which there is more intent on the child's part than actually appears in the forms. Here, too, as in talking, there is a stage at which objects from the surroundings are copied, and soon people and animals are clearly given preference as being something very special.

Representation of objects begins at the end of the second or the beginning of the third year and, for most children, soon becomes quite compelling. Active striving to copy lasts a few more years before it drops off abruptly in most children and is gone entirely before puberty, by the eighth to the tenth years. It is probably appropriate to recall here that the time when the urge to represent things is active is also the great period of brain formation, that during this time, the brain reaches about 90 percent of its mass.

If we survey once more the amount of material to be acquired, the extent of just the imitation required for understanding the most significant part of language and the most important environmental relationships, and for fixing what is learned so that it continues to be useful, the duration of the growth processes in man appears in its most important light and in its true significance.

The biologist who approaches human ontogeny from the point of view of research of animalian development will usually understand the rich developmental events so essential for the growth of an individual in a relatively vague way, based only on his everyday experience. He will be little aware of the unusual nature of these events, and only seldom will he appreciate them as developmental steps, as necessary for the attainment of a final form; many of the events, and not the least important of them, he will not consider at all due to the dominance of the morphological and physiological view. This is also connected to the fact that biologists are primarily interested in the early stages of ontogeny, in fetal stages, much more than in any later ones. The view, in this mode of thinking, that special characteristics that are actually human do not arise until late, subsequent to the preliminary "animalian" stages, is quite a handicap. Sometimes biologists who study developmental phenomena have scarcely any interest in those stages at which it seems that all the organs are unmistakably present and in position, and that the only changes taking place are in the proportions of the parts. On the other hand, the interest of psychologists is directed toward very late stages, for only then do expressions of mental activity appear in any variety. Since psychologists are little concerned with somatic changes, the areas of focus for morphology and psychology are not only far apart in time, but produce very different kinds of results.

Our own research has for years been directed toward clarification of late developmental processes, particularly in birds. Excursions into the realm of mammalian ontogeny have confirmed what we have been able to discover in birds: that a complete, broad understanding of the developmental paths of higher animals is only possible through simultaneous consideration of both the structural and mental realms as being two aspects of a single being; and that, furthermore, as morphologists must know the finished form, the final stage, that is the structural goal, so the entire mode of existence of the mature form, including its behavior, must be clear to us if we are to understand the ontogenetic phases. These observations of later developmental phases in birds and mammals then directed our attention to the specialness of human beings, and it is to those earlier studies that many of these fragments owe their origin.

Proceeding from this point of view, let us once again take a close look at the mental developmental steps during the period from the end of the first year to the sixth year. The adoption of the entire linguistic system; the development of the capacity for objective, pictorial representation; the deepening of experience with objects in the surroundings by buttressing the naming function of words: all are steps of broad significance and expressions of a complex course of events in the nervous system. These steps require long periods of repetition, of practicing in ever new, analogous yet divergent situations. It is really difficult to comprehend what it takes in the maturation of neuromuscular organization and in new life situations for just the human hand, through the full development of genetically determined structures and through adaptation to the many varieties of contact with the environment, to become that agent of creativity that it is even in the childhood of each individual—to say nothing of what it becomes in uniquely gifted human beings. We must be able to see all of this in order to realize clearly what long periods of time are required to guarantee the realization of true human potential through the kind of formation that matures as it experiments, in contrast to the largely fixed behavioral modes of higher mammals.

Once the observer has become aware of the necessity for this slow behavioral developmental process of ours, he can see its relationship to other ontogenetic events of structural formation in a light that makes the subject truly clear. How long does it take for a child to develop compassion in its social relationships? How long must the child remain malleable to allow for all the learning processes, all the experiences that finally lead to an ability not merely to form values based on feelings of pleasure or aversion, but to assess actions according to egoism and altruism—the prerequisite for higher ethical behavior?

However, the slow incorporation into one's own experience of unfamiliar ways of evaluation is also a protection for the unfolding individual, personal world experience, a circumstance that is perhaps apt for letting us see that slow development is not to be assessed prematurely as a negative, but above all, as a necessity. Usually, this slowness of our development appears in biological descriptions as a peculiarity of our somatic development to be understood physiologically. More than anything else, certain discharges from the endocrine glands come to mind, and there is even talk sometimes of a "malfunctioning" in inner secretions, which is held to be the original cause of the general slowing down in primate phylogeny. In any event, this slowness appears to be somatic, with many kinds of consequences for mental development; the evolution of the soma appears to be a precondition for the special kind of mental development that is human.

New examination of the realities leads to a concept of the human developmental situation that contradicts such an interpretation. First, it has been discovered that for our special kind of world experience, with its abundance of social relationships and means of communication to be adopted, and with all that our environment has in store for us, it makes sense for the whole organism to have an extended, impressionable juvenile period; and that, therefore, this juvenile period—whose corollary is slow growth—can be understood as a developmental mode that corresponds to our form of existence. The long period of childhood seems not simply a basic, somatic situation but as something utterly in keeping with the world-open existence of humans. This slowness is not the consequence of a system malfunction in an animalian development taken as the norm; nor is it a malfunction that draws our mental behavior away from natural harmony. No, this slowness is built into our ontogeny as a factor in the realization of the final form, just like any other factor of our development.

Let us also not forget that this much-discussed slowness is not the distinguishing characteristic of all of human ontogeny, as is so often maintained; let us recall once more that, in contrast to the way the great apes develop, both the fetal period and the extrauterine spring are conspicuous for rapid growth in mass. Only when these facts are taken into consideration will the singular, broad organization of our development into periods be clearly recognized. The slow period begins as soon as the somatic and mental preconditions for the adoption of diverse social situations have been created, which is to say after the first year of life, when the elements of human posture, speech, and modes of behavior are present. Just as early, rapid growth is related to a sharp increase in brain mass, so is the leisurely growth of the subsequent periods correlated with special aspects of the formation of our mental life.

Pubertal Growth as a Human Peculiarity

IF WE see our developmental phases according to the basic concept outlined here, we shall also look upon a phenomenon such as the growth spurt of humans at puberty in a new light.[20] This late increase in growth intensity will not be considered, as often happens, as an anomaly in comparison with normal animalian growth, as a kind of endocrine malfunction. Rather, we shall take into account the significant inner development of growing humans as well as the unusual way in which sexuality pervades the peculiar human mode of existence. This very increase in growth intensity during a developmental phase in which we might expect to see a slowing down in mass increase analogous to the situation for animals points emphatically to the deeper context within which we must place human developmental periods.

With the growth spurt at puberty come the striking differences in size that are an important aspect of human racial characteristics. The newborns of races of very different sizes and weights are much more similar to each other than the adults are, especially in height.

	Weight	*Length*
JAPANESE	3,031 g	49.3 cm
GERMANS	3,318 g	50.2 cm
NORWEGIANS	3,484 g	50.8 cm

These comparative figures do not reflect the latest increases, but the increases affect all the groups presented here to the same degree.

Differences in height are insignificant until eight years, or even eight and one-half: at that age, European, Japanese, and Sudanese children (the latter grow to be particularly tall) are all about the same size. Not until the growth spurt at puberty does the striking differentiation begin, with results that vary considerably according to race. Thus, until fourteen and one-half years, Japanese children are larger than Occidental children, but stop growing at sixteen and one-half years, if not sooner, whereas for Occidentals, as already mentioned, growth in males continues for several more years after that.

These circumstances deserve increased attention, for they point clearly to a general human commonality in early childhood and youth firmly established in all developmental types of the different races. It would be premature to draw any sort of conclusion about the origins of human races and the relationships among them from these facts, but the information must be taken into consideration as such questions are deliberated, for it is pertinent to our understanding of the various phenomena of growth at puberty.

This very important late growth phase at puberty has also been called the time of the "second metamorphosis," a term that directs attention away from the all-too-exclusive consideration of sexual maturation to the collectivity of phenomena of this period, to all the shifts in physical and mental proportions. During this transformation a change takes place, in form as well as in behavior, that is well suited for making clear that the mind and body are one even during the transformation; namely, the change in total appearance from a predominantly "pycnomorphic" type to an essentially "leptomorphic" type.

It has recently been confirmed again and again that important individual traits of types that research into constitutions has termed thickset (pycnic) and slender (leptosome) appear in the ontogeny of each human being in striking temporal succession. The early period of development, up until years eight to ten, exhibits bodily proportions of an extreme pycnic type, whereas the later phase, from about years ten to sixteen, is more like the leptosome growth mode. This change in form takes place in more or less pronounced fashion in all individuals, no matter whether the adult, mature form tends toward one of the two extremes—thickset or slender—or toward a mode of growth between the two. We (with Conrad 1941) can speak of an early pycnomorphic and a later leptomorphic growth tendency in human development, a tendency that may indeed appear with very different intensity in each individual,

---

Figure 5.2. Gorilla. Every evaluation of the birth state and the infancy of great apes, and of our own early period as well, must begin with the clear fact that all higher mammals pass through the original mammalian birth state while in the womb, that they traverse this altricial phase with closed sensory organs while still in the mother's body, and, with regard to form, that they are born as precocials, no matter how close the bonds with the mother may be. Ape babies demonstrate this by the multiplier factor of their brains (table 1.6) and the proportions of their limbs (figure 1.7). The young gorilla shows clearly this close resemblance to the bodily structure of the adult (also in figure 2.3, where the baby holds itself upright by grasping the mother's fur). The infant also manages very early to hold on to the mother's pelage by itself, as this picture of Jambo, several months old, shows. In chimpanzees, this faculty is even more pronounced. In spite of the fact that apes are precocial in form, they develop relationships with the environment more slowly than do other precocials—ungulates and whales, for instance. However, no great ape goes through a postpartum phase of transformation in which it does not acquire species-specific posture until it is exposed to social contact. Furthermore, no baby ape experiences the slow metamorphosis to world experience as perceived by the mind as we do in that first, developmentally critical year after birth, a stage that marks a clear division of the entire growth period into three distinct parts. We say that this developmental type is a special kind of ontogeny and call it "secondarily altricial." Photograph: P. Steinemann, Zoological Garden, Basel.

Figure 5.2

but whose temporal succession is a general law. [21] With the change in proportions comes a significant transformation in mental activity and its dominant psychical forces. As this occurs, the modes of behavior correspond to the mental features that, according to research into constitutions, are typical for the two extremes of body structure!

The pycnomorphic period is dominated by what could well be termed a fairy-tale kind of experience, which fades away at eight to ten years. During this period, the child's imagination works with relatively few forms according to the laws of the scarcely conscious, dreamy inner life. Concepts change rapidly, images metamorphose easily and are transferred just as easily to the outer world, as possibility. At that age, the emotional life, free from the causal constraints of external reality, rules. Typical are the broad, undifferentiated field of consciousness, the preponderance of complex inner experience, the forcefulness of the emotions. It is just these qualities that also characterize the mature, human pycnomorphic constitution: the preponderance of holistic conceptualizing, the changeable, flexible, very emotional, highly colored experience. Lucky Hans (Hans im Glück*), in the charming old tale, is to me the very image of this "child-man"—just try to imagine this Lucky Hans as anything but a solid, sturdy lad with a radiant, childlike face!

How different is the psychical force that dominates the child's experience after the tenth year, when the pubertal growth spurt begins. At that time, the wealth of phenomena is perceived more sharply in all its multiplicity of detail, and the store of experience grows through impressions. Critical examination; abstract thinking; the grasping of basic principles: all become important as means of mastering the profusion of stimuli, and the intellect as a whole comes to the fore. "Intellectual capacity during puberty is

*"Hans im Glück" is one of the old tales collected by the brothers Jacob and Wilhelm Grimm in their *Kinder- und Hausmärchen* (Folktales for Children and the Home), published in Germany in 1857.——Tr.

usually very good, I would even almost say that one is never again as bright and clever as at that moment of maturation." This slightly ironic statement (Wissler 1943; see note 21) emphasizes the characteristics that appear in very pronounced fashion in the extreme leptosome form variant of the adult human.

Does it not indicate deeply concealed relationships between body form and type of experience that two such striking, externally evident stages of our development show the mental traits that can be ascribed to the same body type in the adult?

Recent Changes in Maturation

SEEKING TO find out the special nature of the relationships appearing here leads to consideration of a phenomenon that is clearly related to the pubertal growth spurt: the increase, substantiated by much research, of the average body size in many countries of the Western world. This statistical observation has attracted much attention.

Once again, we point out only a few, striking facts. It is known that during the last century, the average height in Europe and America has increased by several centimeters. Since most of the statistics have been obtained as a consequence of military conscription, the statement relates primarily to males. The averages for Basel, for instance are as follows:

1888–89	166.4 cm
1908–10	168.9 cm
1927–30	170.3 cm
1940	172.4 cm
1962	173.9 cm

The same situation prevails in all northern and central European countries, in the United States, and in Canada. It can hardly be a question of a change in the human type, as many optimists would perhaps like to believe. The truth is, rather, that the number of men belonging to the taller, more slender form of the white race has increased. The percentages of male residents of Basel over 174 cm tall are as follows:

1927–30	29 %
1939–40	36 %

If we say that even 170 cm is on the tall side, then the increase in percentage of the total number is greater yet. The numbers of male residents of the Netherlands taller than 170 cm are as follows:

1865	24.6 %
1926	27.0 %
1936	67.0 %

Measurements taken of drafted, nineteen-year-old Swiss men show vividly the alteration in body size: the decrease in the percentage of small men and the increase in that of tall ones! Here are the measurements of one hundred 19-year-olds:

	Under 160 cm	*180 cm and Over*
1888–90	25.7 %	1.0 %
1908–10	16.4 %	1.6 %
1952	2.9 %	10.0 %
1957	2.1 %	11.9 %
1962	1.4 %	15.2 %

In Sweden, in 1896, only 49 recruits belonged to the group that was over 190 cm tall, whereas in 1941, the figure was 539. Measurements taken during both world wars show that, during those times, the phenomenon of increased growth abated somewhat, but following each of the two wars there was a rapid resumption of the increase. In a few years, the growth mode of the previous prewar period was attained once more.

If we look further into the complexity of this phenomenon, we discover that growing taller begins early. Data on the increase of the average birth weight in the last decades are presently accepted with hesitation. On the other hand, intensive growth after birth is documented: twelve-month-old babies are today about 1.5–2.0 kg heavier than babies of the same age in 1890. The increase in growth is particularly strong during the first year of life. This type of growth can be verified for North America and some European countries; however, measurements taken in London show the sharpest increase occurring between eight and fourteen years, which corresponds primarily to the increase concentrated around the time of puberty.

Puberty begins somewhat earlier today, but in children who are already larger than children of the same age used to be. A comparison of a few statistics for secondary school students in Hamburg illustrates the difference between 1877 and 1957.

Subjects	*1877*	*1957*
ELEVEN-YEAR-OLDS	135 cm	147 cm
TWELVE-YEAR-OLDS	140 cm	153 cm
THIRTEEN-YEAR-OLDS	143 cm	160 cm

Children used to gain about 8 cm in height in three years; recently, they gained about 13 cm in the same length of time. Today, with regard to growth, children are an average of 2.3 to 3.2 years ahead of children of the same age in 1877. Pubertal growth presently begins about one year earlier than it did in 1877.

When we speak today of increased height, we also think of the earlier maturation of sexual functioning. Sexual maturity occurs about two years earlier now than it did for people who are today in their sixties. Around 1845, in central Europe, the average age of puberty was sixteen to seventeen and a half years, whereas today, it occurs at 13.7 years. We have more precise data from

Hungary thanks to the famous gynecologist Ignaz Semmelweis (1818–65). In 1850, puberty occurred at fifteen through nineteen years. Contemporary figures are about fourteen years for 1930 and twelve years, seven months for 1959. The data are naturally more accurate for females than for males, for the onset of puberty is more conspicuous and easier to notice than the more concealed changes that take place in the male. The world wars also somewhat reduced the advancement of puberty. In 1945, the regression was about one year, and, as with growth, reversed itself relatively rapidly. Sometime around 1952, the earlier maturation of 1938 was reestablished.

At the turn of the century, the phenomenon just described was still clearly an urban situation and was much less in evidence in rural areas. Moreover, recent careful studies (1965) show that this difference continues to exist, although to a much lesser extent. Within cities, the phenomenon was most noticeable in the secondary schools, a sign that social class played a role. Today, this contrast is much less conspicuous that at the turn of the century. On the whole, the socialization problems that early gave rise to the term "awkward age" *(Flegeljahre)* are today on the whole aggravated and, more important, last longer. If you look in old dictionaries and pedagogical texts, you will find the "awkward age" relegated to the twelfth to fifteenth years; today, we would hardly hesitate to declare the behavioral imbalance so described as lasting until the eighteenth year, and often enough, we are tempted to extend this stage of life even more.

At the beginning of our century, optimism reigned as all of the facts just presented were assessed. Belief in progress and pride in better living conditions and in the many victories of medical research furthered this position. The positive interpretations were supported by many doctors. "And so, it becomes one of the doctor's most ideal tasks to give direction and act as guide along the path toward the highest physical beauty." This is the final sentence in an article on growth (Friedenthal 1914). In 1937, an American anthropologist declared that, clearly, "racial improvement" was taking place in the United States. That harmless prophets thought they saw in these phenomena an evolutionary trend toward the better and more beautiful is less to be faulted than the ungrounded optimism of those who should have been able to see through such evidence more clearly. Seitz (1939; see note 21) was correct when he warned against "seeing in increasing height nothing more than a favorable sign for the race." It sounded a little odd to us Europeans when in 1963 a Japanese scientist announced with some pride to German pediatricians that by 1980, the Japanese would be as tall as the Germans were at that moment. As though some racial blemish would be vanquished from the earth!

Many of the diverse events of the last century have been summarized by the word "acceleration": advanced growth, premature sexual maturity, early attainment of final bodily size! Meanwhile, a more circumspect examination of present-day development reminds us that care should be taken in the use of a collective term that is too one-sided.

Thus, it appears clear enough that the emotional and intellectual maturity required for complete integration into the social group has not kept pace. The original developmental situation is characterized by a very close temporal correlation of the onset of sexual function, the essential attainment of intellectual maturity, and the cessation of growth. At the least, the processes overlapped considerably. Until well into the nineteenth century, the earliest onset of puberty took place at the fifteenth and sixteenth years; the appearance of significant emotional-intellectual maturity was at about sixteen to seventeen years. Today, on the contrary, both the great growth spurt and sexual maturation begin in the twelfth year, whereas the personality still does not mature to any degree until the sixteenth to eighteenth years. The result is that the inner readiness for integration into society is often postponed even longer.

These displacements have altered the entire aspect of the transformation period. In earlier times, the temporal coincidence of all the processes often introduced considerable turbulance into the metamorphosis to adulthood and for some, created a definite period of crisis. The developmental years as Stefan Zweig described them in his retrospective *Die Welt von gestern,* and many similar depictions of crises of youth around the turn of the century, do not speak to the most recent younger generation as reflections of their own youth. The present extension of the period of change has made the tensions of crisis somewhat less frequent and has evened out the entire process. And yet, although the violence of many of the phenomena has been tempered, the duration of the metamorphosis has resulted in a long period of imbalance—one of the most difficult, significant problems of our time.

However, a glance into an even more distant past warns us to guard against an over-hasty simplification. An examination of history shows that it is likely that a long phase of intense growth, which in Europe lasted on into the eleventh century, can be distinguished from a period of less growth from the twelfth century on, a new development that extended on down into the nineteenth century and our own culture. When people happen to talk of that period today, they often remark that the old armor is too small for today's adults.

Therefore, the type of growth we have today does not seem to be the result of a unidirectional process. In addition, it has been discovered that the very period when stature was small, in the fifteenth and sixteenth centuries, was a time of early sexual maturity: the onset of sexual maturity was pushed back from the seventeenth to eighteenth years to the fourteenth to sixteenth; the situation did not change until the end of the eighteenth and the beginning of the nineteenth centuries, when a later maturation type appeared, which in turn was followed by a new developmental countercurrent. This means that in the long period between 1100 and 1800, the processes combined today under the rubric "acceleration" were independent of one another.

The attempt to understand the historical transformation in growth and maturation also enables us to find the restraint that is necessary as we question the causes of the contemporary situation. Indeed, we know little that is certain, and many of the multiplicity of suggested explanations are pure supposition, based primarily on preconceived ideas about the good old days and the collapse of society in our own time. Our attempt to survey the problem of cause proceeds in two directions. The first will be primarily a look at questions of genetic events, and the second will be directed more toward the influences of the prevailing environment.

We know that in every human group, as in any population of animals or plants, lasting genetic transformations called mutations take place that influence the entire spectrum of human attributes and, therefore, affect physical and mental-emotional characteristics to an equal degree. And we should at least bear in mind that, in addition to these transformational factors, already understood through genetic research, there must be other events taking place in the protoplasm, events that for the most part, we are still unable to understand substantively.

A widely noted group of theories on the events under discussion here sees a cause of the phenomena residing in an unconscious social selection of mutants. These attempts at explanation take particular note of the social problem of the migration of rural populations to urban centers. One can start with the assumption that the urge to migrate to cities overcomes some genetically predisposed groups more strongly than others. It might be that more flexible, restless people dominate these predisposed groups. These observations, of course, have nothing to do with a value judgment on the nature of this flexibility and of the multiplicity of causes of this urge. It is critical for this proposed interpretation that a large number of a particular constitutional type migrate from rural areas to the city. No matter how different their fates in the city may be, the assumption may be ventured that many of this very type ascend the social ladder and, in the course of time, enlarge the component of intellectually lively, sharp-witted people in the higher levels of society: a component upon which the technological transformation of the Western world has depended.

Along with this line of thought comes the idea that the breed of man just mentioned resembles a type that has been called from a medical point of view vegetative-stigmatic, and that in all these instances a particular body form dominates—the slender leptosome. In this type of growth, certain intellectual components come to the fore in a more pronounced way, disrupting the elementary harmony of mental activity in favor of greater rationality. Such a correlation has something to do with the performance of this type in the urban setting, for professional and academic selection has long operated in favor of intellectuality and rationality.

Meanwhile, the science of genetics has brought up another possibility for sociologists to consider: the "heterosis" observed in the breeding of plants and animals. The term refers to the frequent abundance of hybrids of species whose lineages exhibit considerable differences. The joining of relatively different genotypes inhibits the sort of genetic balance found in a stable population and can favor the appearance of strikingly new phenotypes, among them being increased growth.

In human society there is a factor that operates very specially in this sense: the increase in number—over long periods of time and due to increased international contact—of marriages whose partners come from widely different populations. I do not even have to mention the enormous dislocations of huge numbers of people brought about by the Second World War and subsequent upheavals. Even events much less tragic raise the level of contact between human groups that in earlier times were quite stable. In my country, the percentage of marriages between Swiss men and foreign-born women rose from 5.5 percent between 1886 and 1890 to 14 percent between 1951 and 1955! That wartime brought a regression to 6.5 percent is also significant for our discussion. The reverse combination naturally occurs even more frequently, but less attention is paid to such unions in our country for reasons having to do with citizenship.

Careful examination of marriages in a small German town revealed that, around 1700, the average marriage radius for this community was about thirteen kilometers and, in 1750, was roughly twenty-five kilometers, whereas today the radius has grown to 140 kilometers. Recently, this increasing proportion of distant marriages has been held responsible for the change in male stature. Research in this direction is still in its inception; it is worthwhile to keep such work in mind when trying to understand phenomena as complex as the changes that take place during childhood and adolescence.

The extensive knowledge of inheritance processes brought to us by modern genetics, especially by the enormous developments in research on population during the last decade, may delude many specialists doing local research in this area into considering the problem of changes in growth to be solved, and therefore, into assuming that the most essential processes are more or less understood as genetic events.

Meanwhile, there are attempts, which must be taken at least as seriously, to explain the problem on a completely different basis. In these interpretations, man appears as an individual marked by environmental influences in both his physical and mental characteristics and shaped by factors that are neither transmitted nor inherited. Many scientists have applied themselves to the very comprehensive study of such environmental effects.

Thus, for example, it must be taken into consideration that since the great population increase of the nineteenth century, the nutritional situation in the Western world has changed continuously, and, in fact, in the direction of an increased proportion of high-quality protein in the diet of broad classes of society. It should also be advanced that, at first, the transformation probably affected primarily a privileged segment of the population, whereas in recent decades, large sections of the highly technological Western societies share in the better diet. The gradual expansion of the phenomenon we are discussing here is directly related to these social circumstances. The rational investigation of nutrition; some insight into the importance of vitamins; increased knowledge of the value of different kinds of food; the will to influence body weight, health, and appearance by means of an ever-broader control of nutrition: all these factors have worked in the same way that increased protein intake has. However, we should not overlook the fact that in many localities, unimpeachable observations show that during long periods of economic distress, height continues to increase and early maturation continues to appear in spite of conditions. Observations of this kind have been made during economic crises in Japan as well as in the United States and Czechoslovakia.

We also do not want to slight the fact—which has surely been emphasized for good reason—of the extent to which mental factors play a role in all these processes. It could well be that the regression during war years is due less to the change in nutrition than to the special emotional-mental situation imposed on whole populations by such catastrophes.

Of the environmental influences that we would like to hold responsible for the puzzling phenomena of accelerated development, the changed relationship of people to light is next in line, for, in our century, a totally new relationship to fresh air and sunlight has grown steadily; we could probably even call it secular sun worship. However, when it comes to pointing out the effects of this new light culture in detail, the situation becomes difficult and contradictory. It seems to me that, today, in the general assessment of accelerated development, the craving for light is somewhat devalued as an explanatory factor.

Meanwhile, the complex event of "urbanization" is in the limelight today. The special culture of large cities, once only local, today increasingly affects the entire population. At issue are the results of many subconscious mechanisms using all the paths of the mind to change people. When we speak of the flood of stimuli, we should not only think of the fragments we all absorb consciously, but must give more consideration to the expansion of the totality of stimuli created by modern life, the effect of which we can scarcely evaluate. The changing situation with regard to illumination has extended daytime life on into the night. The natural daily rhythm has been upset in many ways, when not eliminated entirely, and sometimes drugs are used to help ensure the necessary hours of sleeping and waking. The time available for recreational reading and the corresponding increase in reading material are contributing factors. The influence of film, radio, and television, as well as of mass sporting events, but also the increased intellectualization of education: all operate in the same way; namely, with a relatively modest share of conscious experience and a large, unknown charge of the effects of stimuli of which we are unaware. It is no accident that one of the best experts on this situation as it applies to young people, C. Bennholdt-Thomsen (Cologne), describes it as the "trauma of urbanization." It is also understandable that the very scientists and doctors who work with the negative effects of this trauma assign particular weight to Bennholdt-Thomsen's concept.

No matter how little we can say with certainty about the concealed events that determine how the human form develops physically, it now becomes clear that the singularity of our growth is based on the special interweaving of genetic factors and environmental influences that characterizes our postpartum development. Whatever it is that determines the development of our posture and speech and the ways that the social matrix shapes our inherited structures also brings about the changing events of growth and maturation in different eras.

Many signs indicate that mental experience has its place, and a significant one, among these influences. The problems confronted by psychosomatic medicine are also encountered at every turn by the scientist who studies individual development.

How external circumstances put their stamp on physical and mental characteristics is still largely unknown. This makes it all the more important to lay the groundwork for an understanding, to point out, at least, the paths research would follow in the hopes of finding explanations.

We know that our eye and the optical tract of the brain connected to it perform two very different functions. One, which dominates our conscious experience, is the sensory portion, which provides the visual relationship with the environment. In addition to those goings-on of which we become aware, the eye transmits many effects that proceed from it in the same way, producing optical impressions, but of which we remain completely unaware. In addition, however, optical stimulus also produces another important element, which we might call the energetics component. It consists of an unconscious flow of stimuli that is directed by the optical tract primarily to certain zones of the midbrain, the hypothalamus, where it causes the formation of hormones in the nervous system. These neurosecretions release into the bloodstream substances the effects of which we are just beginning to study closely and which operate partially on certain zones of the hypophysis. We already know some of the possible energetics effects. For this discussion it is important to know that, among other things, in experimentation with animals increased growth has been observed that can be attributed to such optical stimuli. Moreover, animal breeders know that under certain circumstances, acoustical stimuli prompt weight gain or increased milk production in domestic animals. And as we ponder the idea of music in the barn and its economic consequences, we come up with some new ideas about the causes of our own growth mode.

The response system of the organs—here again, genetically determined structures—responds to the neurohormones produced, the organs thus becoming accessible to optical stimuli. Therefore, the study of hormonal glands and the synthesis of their hormones using biochemical techniques is only one goal; the second is deeper insight into the genetically determined reactions of the various organs—and, especially, knowledge of the effect of a particular hormone on the same organ at different stages in our lives. An example may make this aspect clearer. For this, we choose an event in which the thyroid gland, so important for life, plays a part.

In connection with the battle against goiter, the effects of increased iodine intake in humans and on overall growth in childhood and youth has been examined. First, it appears that such an increase in the amount of iodine in the mother's body has no influence on the developing child, but also that supplementary iodine has no effect on the child until the seventh year, and that, finally, even the late phase of puberty, when the impulse to growth diminishes, is not influenced by increased iodine content. In contrast, when iodine is given to eleven-year-olds, who are at an age that today is especially important for pubertal growth, a definite increase in size appears. Therefore, an influence such as an increased supply of iodine is fully effective only during a very limited period. It goes without saying that these experiments say nothing particular about the influence of thyroid and iodine metabolism on the events that especially concern us here; primarily, it is a matter of simply noting how strictly the responses of the organs can be delimited temporally. I mention this example because it permits at least a glimpse into the workings of pubertal events and can thus serve as a symbol for unknown influences on the growth period.

While educators study and weight the effects of conscious sensory impressions, especially of pictures and words, the special system that directs sensory stimuli to the neurohormonal tracts is a realm still to be investigated, one that may also contain within it the solutions to the individual riddles of human growth.

Today, we have much evidence to show that the organs of our bodies respond to stimuli given off by our emotional and intellectual experiences. The capillaries of our skin serve not just as thermoregulators for our bodies; they also respond to changes of mood, to emotion, even to intellectualized stimulation, such as feelings of shame. It is conceivable, then, that in the critical years of growth, the production and the control of stimulative substances depend much more on external impressions and emotional stimulation than we at first supposed would be the case with such distinctly "physical" events as the formation of the growing body. It is perhaps not completely in error to assume that the kind of stimulus that operates during the period of juvenile growth is not without influence on the regulation of the mode of growth. Thus, we venture to expand upon the hypothesis of C. Bennholdt-Thomsen on the effects of urbanization by adding an element. We posit not only increased response-readiness in certain people, but, in addition, a selective accumulation of the mental stimuli to which the disposition to slender, leptomorphic growth responds particularly strongly. We think that in daily urban life, and to a surprisingly increased extent during the most recent decades, a kind of intellectual activity dominates that we find particularly characteristic of the extreme leptomorphic type. This one-sided increase of certain portions of mental output is undisputed—although it may not seem that way from the many complaints one always hears about grades at school!

That correlation already pointed out between physical form and mental activity exists not only in the extremes of constitutional types; that it also appears in a similar fashion in the two phases of the developmental period; and that, finally, the one side of the correlation (intellectualization and increase in height) appears again in our schools: all of this should scarcely be regarded as an accident or a coincidence.

Perhaps the so conspicuously enhanced growth in height (we use the most striking trait first) is triggered by a particularly one-sided enhancement of certain mental activities during childhood and adolescence, an enhancement that occurs during the period of formal education, in general, assuming unusual proportions in secondary schools, but one that is also favored by the general orientation toward intellectual activity of our entire age. If such one-sided augmentation continuously strengthened that growth impulse which determines the leptosome growth mode, the result would be changes in proportion at the critical age in every individual, in varying degrees according to his predisposition. In potential pycnomorphs, such impulses lead to slight changes, but affect constitutions that are more leptomorphic to a larger extent, so that as the leptomorph pole of form is approached, enhancement of growth in height and of many developmental accelerations is considerably increased. This preliminary attempt at interpretation also takes into consideration the fact that we are not talking about an increase in size of any particular racial type, but about an increase in the number of slender to very slender individuals within a population—a shifting of the average values toward the leptomorphic somatic pole.

Our interpretation is a working hypothesis, which must be tested by new series of observations. However, it seems to us time for the much-discussed phenomenon to be observed by a new method, one with the potential for yielding an understanding of many peculiarities of pubertal growth. At issue are basic questions of cultural influence and the possible control exercised by such effects. The dominant view of these matters used to be largely optimistic, and it is not our intention to simply substitute a pessimistic one; the task of biology must be to understand the reality of this phenomenon in its totality.

"As matters have developed, the preponderance of our faculty for understanding has grown out of disciplined mental activity, its weight seen written beneath hard foreheads in the empty faces of mankind as it scurries about, and so it has happened that today, smartness and stupidity—as if it could not be otherwise—are perceived simply in terms of the intellect and how proficient it is, although this is more or less one-sided." That the bias Robert Musil complains of with these words[22] is one of the ominous manifestations of our time must be said over and over again, and in view of the significance of this question, it should probably be required that the connection assumed here between modern man's intellect and his outward appearance be examined seriously.

The phenomenon of increased height, which we do not regard in an optimistic light, is for us one of those cracks in a firmly assembled whole through which shines something concealed in the depths, and we observe it carefully because of the glimpse it might provide of the inner members of a complex structure. We have every reason to pursue childhood and adolescent development objectively. These days the matter is only too often assessed at a level that is not far from the ideal of beauty found in the movie world and fashion magazines; the danger hidden by such an approach is that it will completely cloud our view of the reality of our form and being.

6
Senescence

IT IS one of the givens of biological observation that in age, as human beings wither and fade, the correspondence between our existence and the lives of animals is extremely clear and convincing. Images of shriveling and decay, of declining strength, are ever regarded as expressing the human being's inclusion in the animal kingdom. Even the emphasized and exaggerated lifespan of many animals has lent support to the idea that our own span of existence corresponds completely to the potential for animals, that it is one of the strong links our species has with the general laws of animalian life. And yet—no matter how obvious many of these symptoms of old age may be, how much they invite such an identification—the last phase of human life is characterized by peculiar traits that have been too little noticed.

Biological Comparison

THE COMPARISON with the animal has to clear two hurdles. The first: the many uncritical data on the ages of animals, which only humor the human need to imagine extremes. Almost all of the ancient carp, turtles, and crocodiles succumb to an objective examination of these data, as do parrots and elephants, said to live more than one hundred years. It would be worth doing a psychological study to investigate the reasons that move us to elevate certain animals particularly often to the rank of venerable patriarch. Such a study would probably teach us a great deal more about human beings than about animals.

The second hurdle is a tradition of biological comparison that indiscriminately assesses animal life according to absolute time. This absolute scale—after having been determined with absolute precision—must first be subjected to differential evaluation. The lifespan of a simple structure such as a growth of coral should not be equated with the same time span in human life; a hundred coral-years must be appraised differently than a century of vigorously lived human life. The fact that, in design, the two beings have very different structural complications is not the only constraint; the problem is that the coral growth taken as a whole is not in the least comparable with an individual human being. The concept of the individual, applicable to the higher animals, has just as little validity in the realm of these polyps as does the concept of individual death.

The difference in the temperatures at which the life processes are carried on also militates against equating the two types of beings. For corals of the tropical seas, the ambient temperature of somewhat above twenty degrees Celsius determines the intensity of metabolism; our own body temperature of thirty-seven degrees Celsius supports a completely different rhythm of existence. The turtle's year, half of which, in our geographic zone, is spent in a completely lethargical winter hibernation, must be judged differently by biologists than that of the tern, which twice a year flies almost from pole to pole, from the northern polar summer to the other summer in the Antarctic.

Making such an appraisal is not a matter of subjective opinion; that is, it is not a matter for discussion whether the individual human who knows of this difference in animalian life gives preference to the unconscious, slumbering existence of the corals over the swifter rush toward death of the warm-blooded creatures. We are simply saying that more activity takes place in one life than in another, that one life can have more effect on its environment than another, just as we say that one thousand is more than ten, which has nothing to do with whether we consider a life of poverty or one of wealth to be the most fulfilling human existence. That is why we take only warm-blooded animals into consideration for our comparison, and even from this category we shall make choices: only with caution, for example, including in our discussion animals that hibernate. It takes only a few figures to summarize what should be regarded as being relatively certain about the lifespan of mammals.

Laboratory animals have been studied in the most detail. The white mouse and the white laboratory rat reach an age of two to three years; variations in diet can slightly lengthen or shorten this span. Guinea pigs live six to eight years, rabbits ten to fifteen. The typical lifespan for dogs can be set at twelve years, for cats at sixteen, and for horses at twenty. Large predators and ungulates live to be thirty years old. The age limit for large whales is thirty to thirty-five years. Precise investigation has shown again in 1967 that elephants, both the African and the Indian, reach an age of twenty-six to twenty-eight years on the average; higher estimates give thirty years, rarely forty.

For birds, in individual instances advanced ages of sixty and seventy years have been reported; in mammals, only a few exceptional individuals live more than fifty years, and even those do not exceed the half-century mark by much.

The singularity of our human life stands out even more vividly when compared with that of the great apes which, in any event, are the most important comparative element: orangutan, gorilla, and chimpanzee age very early and are senescent at between twenty and thirty years. Very few are likely to live beyond the third decade. Riopelle (in Schreier et al., vol. 1; see note 3) points out that the skeleton of a forty-year-old chimpanzee was in a similar state to that found in a human being of between sixty and eighty years of age. Yerkes (1939), with all due restraint, emphasizes that twenty-year-old chimpanzees appear to be as old in mind and behavior as we do at forty to fifty years. This is what is critical for our discussion. Yerkes also stressed the early, "adult" seriousness of the chimpanzee, the complete disappearance of play at about ten years, also a sign of the different tempo of that life.

Assessment of the life span of anthropoids in captivity must naturally take into consideration that many of the animals die prematurely of infection. On the other hand, the ever more frequent reproduction of great apes in zoos proves that their living conditions have become more favorable. We can probably assume that, today, animals reach their true age limit more often in captivity than in the wild, which means that it is safe to infer the possible lifespans of animals from the more recent data provided by zoos.

Just the comparison of our lifespan and senescence with that of animals of our own structural type shows the special nature of human life. We can probably compare the presumed lifespan for great apes—about thirty years—with a human age of sixty to seventy years. Our existence is distinguished from that of apes by a lifespan about twice as long. The same situation is valid for the extremes. If, on rare occasions, an ape should surpass the fourth decade, its age would correspond to the advanced human age of eighty to one hundred years and above. There is no credible evidence of such an advanced age in mammals other than man. Thus, a purely quantitative assessment leaves man occupying a special position among the mammals, one that stands out vividly in comparison with that of the animals most closely related to him. But observation based on raw statistics prompts a careful, more general investigation of the phases of human aging. When the thoughts set out in this book were first presented as lectures in 1939–40, I had to emphasize how unsatisfactory the descriptions of human aging and senescence still were. Since the appearance of studies by A. L. Vischer, which have led to a more profound consideration of these life phases, research on old age has become a separate branch of science–gerontology.[23]

More recent investigations have called attention to the problems facing research on comparative senescence. Between 1957 and

1962, Suzanne Bloch studied the aging of sexual function in female rats, comparing the results with the situation in humans. Biologists come to the conclusion that one rat year is roughly equal to thirty years of our own lifespan. The juvenile stage of rats, attained at about three months, is the equivalent of our maturation period of about fifteen years; the advancement of sexual maturity in recent decades has led to an even closer correlation! Even the period of time during which the organism is fertile (twelve months for rats, about thirty years for women), is strikingly similar in its relationship to the total duration of life. After cessation of genital function, rats go on to live about another twenty-one months, which corresponds to about the ninetieth human year.

These studies invite careful examination of other mammals: in particular, of apes and great apes. At issue is an important question of our phylogeny, of the problem of how the specific prolongation of human life came about. Is this long period of senescence a late-arising "luxury" for our form of existence, as has often been thought, or does the potential for advanced age already exist in mammalian genes—Suzanne Bloch's view—and can it, under favorable conditions, lead to a special senescent phase? In any event, the comparisons we have made here show how fruitful it is to consider the entire course of an individual life as a whole, and to attempt to place the individual elements in context.

Human Singularity

QUITE OFTEN, biological descriptions of human senescence emphasize exclusively those traits that typify the gradual decline from the peak of physical fitness and the various mental traits easily associated with that decline. Even when it is conceded that a human's life curve is longer than that of most other mammals, it still appears in biological research as just an insignificant modification or variant of an animalian lifespan. Certainly, there are instances of human decline in old age that are similar to events in animals' lives—but should they really be considered the norm? As soon as we begin to observe human existence seriously and are willing to comprehend it in its entirety, traits and images will appear, singly at first, then with increasing diversity, that are completely different, far from anything we have observed in animalian life. The venerable figures of the ancient prophet and the wise old woman—once these revered images are conjured up, many related forms of humanity at its highest pass in review before our inner eye. We see the mighty, legendary visionaries and sybils, so many of whom are distinguished by the features of great age. The conviction that the aged are more powerful intellectually continues to be a vital one for many different peoples. We encounter oligarchies of the aged as a significant social force in the aborigines of Australia, and elsewhere, among other peoples, we find widespread great respect not only for the advice and counsel of the aged but for their literary testimony as well. Indeed, this conviction is so strong, the language of experience so powerful, that the creative imagination has also endowed old animals with special wisdom.

Even though human senescence is similar in many respects to that of animals, a completely different potential has been granted us. No matter how accurate the image of decline to a childish state may be, the power and truth of that other image, that of the special worth of advanced age, stands parallel to it, demonstrating the singular significance of this phase of advanced age in a being whose very existence is based on the transmission of the great riches of experience and culture. This is what is so peculiar about the course of our lives: every degree is present, from animalian senescent decline to the moving, late creative power of many human beings. As long as the only characteristic of old age that is taken into consideration is the downward curve, the wilting of the flower of life, and not the ripening of the fruit, the new, imperative problems posed to society by old age will be conceived incorrectly and solved incorrectly. The prerequisite for a promising concept of the new task facing society is the clear recognition of the singularity of human old age.

As we age, one of the characteristics of our peculiarity in general expresses itself with great force—heightened individuality, the mark that sets us apart. It is just this phenomenon that biological methods of observation very often deny, for biologists, trained to conduct research on plants and animals, place particular emphasis on what is typical, on what is held in common. This position has also led research to approach the individual by way of type of temperament, constitution, or character, most often by summarizing recurring combinations of characteristics. One could hope to have at least some aspects of a characteristic as prominent as human individuality respond to research into general principles. Even those who recognize the value of attempts to construct a typology will be in favor of a nuanced consideration of individuality, for the mood of the time tends to join battle against "individualism" and confuses this objective all too easily with the challenge to the worth of the individual as such. Before we unhesitatingly see typology as progress, we must at least answer for the fact that it contributes to one of the gravest, most momentous of contemporary processes: the devaluation of the individual.

More extensive, detailed consideration of the phenomenon of aging may show particularly vividly the broad range of individual developmental differences: one of the great, fundamental facts of human society and, therefore, of all biological consideration of human existence. The destinies of the greatest creative energies reveal these contrasts between different lives with particular clarity, but as part of a vast series of individual destinies marked by various levels of intelligence, they are lumped in with those many lives that are called "normal" because the latter are in the majority. Just the consideration of the last phase of senescence indicates to us how important it is to acknowledge, in addition to such purely quantitatively determined norms, the significance of deviations. The alternative is unwittingly to make out of the mass-deduced norm a kind of abstract picture of existence in which a judgment about the frequency of a phenomenon is mixed up with an evaluation of its worth.

It is high time that the peculiarity of human old age is completely understood, for progress in the fight against sickness and accident, progress that remains an important, positive fact even though "modern technology" is regarded with pessimism, has led to the fact that, in the future, an even greater number of people will live longer. The increase of the average life expectancy from about forty-five years in the last century to about sixty-four to sixty-five years and more has led to many social problems that most people—particularly those called upon to lead their countries—are still at a loss to solve. This perplexity is only poorly concealed by emotional talk about the danger of a society's becoming superannuated and senile. If an aging person were nothing more than a being in slow decline, as many biological considerations and political statements would have it, the role of aging in contemporary society would certainly not have become as problematic as it is today, and the threat of an entire people's becoming superannuated, painted so blackly, would consist more of just the difficult practical tasks of caring for and maintaining the good health of the aged, who, in retirement, are dependent on charity more or less willingly given.

This catchword, "superannuation," gives only the negative side of a meaningful reality, the positive side of which—the large, often irreplaceable achievements of advanced age—must be brought to the fore. Let us not forget that Sophocles wrote powerful plays when he was ninety years old, that Radetzky* was eighty-two at the victory of Custoza, that Moltke† won the Franco-Prussian war at seventy—and that Titian completed his most compelling works when he was nearing one hundred. If this positive aspect did not

*Joseph, Count Radetsky von Radetz, 1766–1858. Austrian military reformer, veteran of the Napoleonic Wars; as field marshall, he conducted operations against the Italians during the Italian Wars of Independence, which resulted in an Austrian victory at the first battle of Custoza, in 1848.——Trans.

†Helmuth, Count von Moltke, 1800–91. Chief of the Prussian General Staff from 1858 to 1888, during the Bismarck era. He was the architect of the Prussian victory over France in 1871, and also a writer, sometimes considered one of the masters of nineteenth-century German prose.——Trans.

exist, young people today would probably not have to be so vehement in their demands for the aged to retire. Societal traditions would have long since found a solution to the problem; it would have been settled "biologically" if, that is, biological insight were really so unequivocal and our senescence really nothing but the more or less swift decline from life's climax. A clear view, one firmly rooted in complete insight into the true riches of the problem posed by the increasing proportion of aged people in the total population, is vital for the intellectual leadership of every land. It is also the prerequisite for the productive functioning of democratic institutions. And the way toward this goal should not be barred by fundamental biological premises based exclusively on animals and by a medical science that operates using biology's concepts.

Contemporary biological knowledge does not yet have an explanation for the peculiarity of human senescence. We still lack knowledge of the details of the aging process. But this very period of aging directs the inquiring glance again and again to what we have already mentioned several times in this investigation: to the relative dominance of the central nervous system, the singular influence of the highest mental activity on all other bodily functions.

A glance at the richness of the late phases of human life, at the great differences in vital powers perceptible at just this time, compels us to realize the limited validity of norms established simply quantitatively and to open our awareness to the mysterious heirarchy of the power of the mind.

7
Conclusion

TO SHOW what links us humans to all that lives: many people feel this is the proper contribution of biological research to the formation of a world view. Based on this way of looking at things, the life sciences have explored a kind of vital substructure of all that is human, upon which the workings of the mind then build. Connected with this endeavor is often the hope that this mental activity might be found attributable to elementary life functions.

Change in Point of View

IF WE regard biology's mission in this way, the narrower the gap between human and animal appears to be, the easier the solution, and therefore, the greater the effort to see what is human as insignificant and animalian. Many have gone along with this attempt; in many instances the consequence has been the actual devaluation of human worth and dignity. We probably do not have to go out of our way to find contemporary examples of this degradation.

Acceptance of the belief that the difference between man and animal is slight is facilitated by the fact that the most superficial concepts formulated by theories of the origin of species have become taken for granted by many people. Perhaps the results of our studies present the problems of origin in such a way that their difficulty is perceived once more, showing the vastness of the realm that a scientific theory of origin must encompass. In recent decades, behavioral research has opened up a new view of animal life, one that has brought a high regard for the animalian relationship with the world while deepening our knowledge of the fundamentals of life and behavior that are shared by human and animal. But it is just these significant insights that also sharpen our eye to the peculiarity of our position in the world. The special objective of our work, the emphasizing of what is distinctively human by means of biological studies, arises from the conviction that a pronounced shift in accent, a different cast of light, would render much visible that today, due to one-sided attention to similarities, still lies deep in shadow.

A change in perspective has also become an urgent necessity in the struggle against the misuse of biological concepts as political slogans, as weapons in the internecine wars among human groups. It was imperative even years ago to fight against these slogans, when the struggle for existence was palmed off on man as a creative force; and it is even more important today, when people are trying to construe the entity of the State as an organism—as a being that has survived—talking us into believing in this invention with scientific arguments.

It is to be hoped that the time is not far off when all the commonalities that connect humans, animals, and plants will be as well known and obvious as many basic facts about the structure of the solar system are today. At that point, biological research will be able to do something more than reiterate the proof of this commonality, on which so much effort must still be expended today. Another task will then take on importance: proving the separateness and distinctness of individual organic forms, thereby proving the singularity of human existence. Careful study of all human peculiarities will make it clear once more what a theory of the origin of the human species actually has to offer. Not until a new method of examination gives the human peculiarity its full weight will the mission of a theory of anthropogenesis be seen.

Indeed, until now, the weight given the facts has been very strongly determined by how useful they were as arguments for the theory of descent.

The position taken by our way of looking at things does not undermine scientific research into a theory of the origin of species. In fact, from our position there is a sharper view of the many transgressions committed by a biological science operating far afield of its scientific origins, of a discipline called upon only too frequently to take on problems for which the dwindling faith in ancient religious tradition demands new solutions.

The stages in which the geological phases of human evolution are presented today, almost everywhere and to larger and larger audiences, are powerfully impressive through their use of vivid illustrations and antiquated three-dimensional models. The forceful, expressive quality of artistic representation, the visual power of the scenic reconstructions in large museums, project the certainty that research into these distant events is possessed of complete clarity. But where is the broader representation of the doubt, of the contrasts and contradictions in interpretation such as are encountered in the literature by anyone who really consults and questions the sources? One would do well not to underestimate the skillful persuasion exerted by the consummate realism of presentations in textbook illustrations, charts, and public exhibitions, and indeed in films. The immediate sensory effect of such vivid pictures on people who are already accustomed to the surrogate reality paraded before them by the dream factories of the film industry, on people whose spirits must live from such indirect experience: these sensory effects all too easily enable us to forget that what we are seeing is the expression of plausible opinions and only to a very limited extent the presentation of verified facts.

Such images had already come to the aid of politicized biology before there was ever any ministry of propaganda. Are the biologists who are entrusted with preparing the representations in films and museums ever aware of the responsibility they assume through their works?

In recent decades, research into our social existence has expanded in two ways that are significant for the concept of the human being: the extension back in time of the historically comprehensible past is one; the glimpse into the diversity to be found in the group life of higher animals is the other.

Discoveries of prehistoric material appear increasingly today as historical documents, traces reaching far back in time, showing ever more clearly something peculiar, something human, be it only the signs of the fire that had charred the remains of some object, or the ever so slight mark of the craftsman's hand on the tool of bone or stone: in these signs we always divine the whole human being at work. In many places, and with increasing frequency, collections of prehistoric material are finding their way into historical museums after having spent decades as the culminating pieces of natural history collections. It is an uneventful, seemingly harmless process, but one that speaks volumes. Zones of human prehistory that once informed us mainly through skeletal remains and which, consequently, were left to the methods of biological research are now considered in a historical light, no matter how tentative our statements about the possible life-form of that ancestral group may be.

But even when the biologist reports on higher animalian life, we encounter an element that is familiar to us from human historicity, without which we would no longer regard a description of the social behavior of animals as adequate: the fact of tradition, of a special system of transfer for each group, be it of gestures or of forms of behavior, qualities that are not part of the actual genotype of a related group but which have been acquired and fixed through long familiarity among members of the same species living together. It goes so far that even a behavior that we consider without hesitation to be genetically determined, such as care of the young, does not become stabilized in higher apes until it has been made familiar through years of communal living. The difficulties that young pongid mothers have raising their offspring are particularly great (and can lead to complete rejection) if the mother has grown up alone in a zoo. For adequate behavior to develop—this goes far beyond the events of reproduction—it is critical that the years preceding sexual maturity be spent with the group.

Examples of the workings of authentic tradition through imitation and habituation are numerous in the realm of mammals and birds. We have every reason to assume that during the unknown preliminary stages of the human race, this prerequisite for the special historical human form of existence was operating to a considerable extent.

Insight into the significance of tradition in the lives of the highest groups of animals builds many bridges toward understanding the prerequisites that had to be fulfilled before anthropogenesis could actually take place: long living together of several age-groups, with many-sided social contact, as well as a considerable openness of the hereditary systems that regulate relationships. This structural openness is not to be confused with the world-openness of the fully formed human mode of existence. Nor is tradition as an element of group life to be confused with historicity. On the contrary, the fact of similar modes of communication compels us, because of many suggestive similarities, to construe the peculiarity of conducting life in a historical way, as humans do, more narrowly than ever as we search for traces of the possible origin of this singular characteristic.

Language, with its great, special capacity for abstracting from every situation, every mood, every surge of affect, remains the critical factor for delimiting the animalian realm and for an understanding of how and why we are different. The fact of controlled expression, in contrast to the spontaneous expression of animals and many similar forms of expression in humans, too, cannot be emphasized sufficiently.

There are particular physical forms of expression that testify to the peculiarity of our life-form and are part of an intermediate realm: the play of the capillaries of the skin in blushing and turning pale is closer to spontaneous expression and yet, when feelings of shame are aroused, this same activity is at the service of mental experience in the world. Then, there are the paired expressions of laughing and crying; not without reason are they the cause of so many conflicts about the nature of human beings. These expressions are clearly more like those to be found in the sphere of controlled expression, without, however, belonging to it in the same sense that verbal or gestural language do. A powerful effort is expended during our education and upbringing in the battle against spontaneous expression, a battle that tries to force this mode of communication to serve the controlled relationship. How easily this educational goal changes as generations go by, how swiftly upbringing and education can reevaluate these intermediate forms of expression, now letting the tears flow freely, now controlling all weeping, using the smile to achieve social ends, suppressing the liberating laugh.

Humankind: an Evolving Image

LIFE DEMANDS more than the modest certainty that factual research can offer us. From the scientist who shows others the fragment because even he can unearth only that torso from the ground of experience, from him these others demand a whole.

However, this whole, which must be wrested from the unknown and the unspoken, will never be delivered by research alone; it will arise only as a result of many human mental powers working together. All creative genius, musing, dreaming, going beyond the experiential, is at work on the problem of what is man —the powers of artistic and, especially, of religious experience no less than the powers of pure reason. From such collaboration will arise images of humankind in which is inherent the guiding force, which, through its consummate greatness and power to impress, will decisively influence the deeds of the individual, and by extension, the deeds of the many.

Not one of the images of humankind in effect today will survive the difficult intellectual and spiritual battles of our time unscathed! The future image will have to overcome not only the ancient myths of creation but the intoxication of undisciplined evolutionary thought as well.

New results of scientific research drive human thought relentlessly on, like a wheel that rolls forward propelled by an unseen force. Let us stay with this image for a moment; it also reminds us that each point of this rolling wheel is constantly moving backward as well, without this change in direction being a return to a former position. The work of biology moves in a similar fashion.

The coming insight into the distinctiveness of mankind is a turnabout from the period during which the commonalities of all life-forms have been a particularly strong presence in the consciousness of the times. But it is also a change in direction that we now see clearly once more how really human is the horrifying brutality of our deeds. This evil no longer seems nothing more than the remnant of our animalian heritage yet to be overcome, as the optimistic early stages of the theory of the origin of species would have it, but as the heavy burden of being human. Today, we see through the self-deception that spoke of "reversion to primitive barbarism and brutality" and know that behavioral extremes still exist, their potential slumbering within each of us.

The more clearly the human form of existence appears to us, the graver is the certainty that the means available to research today will not be able to answer the question of human origin or the one —just as difficult—of the origin of the great groups of life-forms. Currently, we are getting a grasp of the origin of many variant forms; the theory of mutation offers an understanding of broad areas of transformation in animal and plant forms and of extensive sequences of form over geologic time. Yet these theories provide no information about the origins of the great groups of organisms, which are indeed the most significant and remarkable events; only by stretching their application do they provide a pretense of an answer to these extremely difficult questions facing the life sciences.

Today there is widespread claim to superficial knowledge of origins; but a new spirit, calm and serious, should come to the fore, one that acknowledges the immensity of the mystery. Against this dark background, the human image will appear, not as the self-assured being from the ancient myths, which assumed that earth and man were the center of the universe—nor yet as that oversimplified image of the amoeba that has made its way up in the world! The transformation that began centuries ago with the Copernican revolution, which in its own day compelled men to see through the delusion that they were the center of the universe and required them to renounce outward appearances, will have to continue on its laborious course, with all attendant inner turmoil. This asceticism will be intensified when thoughts of the limited possibilities of life on this earth, as it presents itself to our senses, are necessarily measured against the boundless possibilities for life in the extraterrestrial universe. The possibility that extraterrestrial life exists in profusion will have to become a continuous presence in our imagination.

Such expansion into cosmic realms is "of-this-world"; it carries biologically based concepts out into the unvierse, where our thought and inquiry have already long been at work. It is just this "of-this-worldness" that gives us the right to speak of such expansion of the imagination in a biological study concerned with the limits of research into life. The wealth of the mind as revealed by research into terrestrial life will only open out to its fullest when we place the fragment of life we know within the mighty context of the vital aliveness—closed to our senses but suspected nonetheless—of the immensity beyond.

To the initiate, deeper knowledge of the order of terrestrial life cannot possibly be a symbol of disorder, or chaos. Insight into ordered events, transmitted by all life sciences in profusion to the point of excess, can only lead to the intimation of still greater order and awaken the feeling for the immensity of the secret, for premonitions that make the darkness of the hidden causes deeper and more pregnant with meaning. The well-seen forms that live around us are evidence of forms and designs greater than what the visible world offers.

Notes and Commentary

1. Detailed presentations of individual questions bearing on the discussion of the problems of evolution can be found in other of my publications:

Portmann, A. *Einführung in die vergleichende Morphologie der Wirbeltiere.* 3d ed., considerably revised and expanded. Basel: B. Schwabe, 1965.

——Zur Philosophie des Lebendigen. In F. Heinemann, ed., *Die Philosophie im 20. Jahrhundert,* 410–40. Stuttgart: E. Klett, 1959.

——*Neue Wege der Biologie.* Munich: Piper, 1961; *New Paths in Biology.* Arnold J. Pomerans, trans. New York: Harper & Row, 1964.

——*Biologie und Geist.* Zurich: 1956.

——*Aufbruch der Lebensforschung.* Zurich: Rhein-Verlag, 1965.

——*Das Tier als soziales Wesen.* Zurich: Rhein-Verlag, 1953; *Animals as Social Beings.* Translated by Oliver Coburn. New York: Harper & Row, 1964.

2. Here, we touch on the controversial question of the interpretation of the earliest prehistoric finds. It is presented in detail in the monograph by E. Baechler, *Das alpine Paläolithikum* (Basel: 1940). Whoever would like to become familiar with the entire extent of the dispute should also read the study by F. E. Koby, Les soi-disants instruments osseux du paléolithique alpin et le charriage à sec des os d'ours de cavernes, *Verh. naturforsch. Ges. Basel* (1942–43), vol. 54. A response to the work just cited: H. Baechler, Altsteinzeitliche Knochenwerkzeuge oder Bärenschliffe? *Jahrb. Schweiz. Ges. f. Urgeschichte* (1943), vol. 34.

On the evolution of artistic expression: W. Matthes, Die Entdeckung der Kunst des älteren und mittleren Paläolithikums in Norddeutschland. *Ipek (Jahrbuch für prähistorische und ethnographische Kunst)* (Berlin: 1964–65); vol. 21, and Matthes, Zum Verständnis der älteren Eiszeitkunst, *Antaios* (Stuttgart: 1967), vol. 9, no. 3.

3. For information on the present state of our knowledge of anthro-

poid apes, see first R. M. Yerkes, and A. Yerkes, *The Great Apes* (New Haven: Yale University Press, 1929). Then, R. M. Yerkes, *Chimpanzees, a Laboratory Colony,* 4th ed. (New Haven: Yale University Press, 1948); A. H. Riesen, and E. F. Kinder, *The Postural Development of Infant Chimpanzees* (New Haven: Yale University Press, 1952). Several studies by G. Brandes appear in the journal *Der Zoologische Garten,* new series, 1928 to 1935. In addition:

Brandes, G. *"Buschi," ein Orangkind.* Leipzig: Quelle, 1939.

De Vore, I. *Primate Behavior: Field Studies of Monkeys and Apes.* Cambridge: Harvard University Press, 1965.

Hofer, H., A. Schultz, and D. Starck. *Primatologia.* Handbuch der Primatenkunde. Basel/New York, since 1956.

Kohts, H. La conduite du petit chimpanzé et de l'enfant de l'homme. *J. Psychol.* (1937), 34: 494–531. (Observations in Moscow with very careful studies of vocal expressions.)

Kummer, H. *Social Organization of Hamadryas Baboons.* Basel and New York: Karger, 1968.

Lang, E. *Goma, das Gorillakind.* Zurich: A. Müller, 1961; *Goma, the Baby Gorilla.* E. Fischer and C. Schoffer, trans. Garden City, N.Y.: Doubleday, 1963.

Nissen, H. W. *A Field Study of the Chimpanzee.* Comp. Psychol. Monographs (Baltimore: 1931), no. 8.

Rensch, B. *Handgebrauch und Verständigung bei Affen und Frühmenschen.* Symposium of the "Werner Reimers-Stiftung für anthropogenetische Forschung." Bern: 1967.

Rode, P. Comparaison entre les Anthropoïdes et l'Homme au point de vue du cycle génitale et du développement des jeunes. *Bull. Soc. Anthropol.* (Paris: 1939), series 8, vol. 10.

Schaller, G. *The Mountain Gorilla.* Chicago: University of Chicago Press, 1963.

Schrier, A. M., H. F. Harlow, and F. Stollnitz. *Behavior of Nonhuman Primates.* 2 vols. New York: Academic Press, 1965.

Steinbachner, G. Geburt und Kindheit eines Schimpansen. *Z. f. Tierpsychol.* (1941), vol. 4 (Observations made at the Frankfurt a. M. zoo; many important details.)

4. A particularly valuable introduction to the emotional life of children is offered in the works of the pediatrician F. Stirnimann, of Lucerne; these studies have inspired our own in an important way. Its list of recommended books and articles guides the reader further into this field of study. F. Stirnimann, *Psychologie des neugeborenen Kindes* (Zurich/Leipzig:

Rascher Verlag, 1940); and, from the same author, *Das erste Erleben des Kindes* (Frauenfeld/Leipzig: 1933). The detailed study by Brock is indispensable: J. Brock, *Biologische Daten für den Kinderarzt,* 2d ed. (Berlin: Springer, 1954).

Psychology, influenced by psychiatry, has made important contributions to the knowledge of mental development:

Christoffel, H. Einige fötale und frühstkindliche Verhaltensweisen, *Int. Z. f. Psychoanalyse und Imago* (1940), vol. 24, no. 4.

——*Skizzen zur menschlichen Entwicklungspsychologie*. Aarau: 1945; 2d ed., Bern: Huber, 1965.

——Über-Ich und Individuation. *Schweiz. Z. Psychol.* (1947), vol. 6, no. 4.

Fordham, M. *The Life of Childhood.* 3d ed. London: 1944.

Langeveld, M. J. *Die Schule als Weg des Kindes*. Braunschweig: G. Westermann, 1960).

——Studien zur Anthropologie des Kindes. Tübingen: M. Niemeyer, 1956.

Spitz, R. A. Hospitalism: An Inquiry into the Genesis of Psychiatric Conditions in Early Childhood. *The Psychoanalytic Study of the Child* (1945), vol. 1.

——Analytic Depression. *The Psychoanalytic Study of the Child* (1945), vol. 2.

——Emotional Growth in the First year. *Child Study* (1947).

——La perte de la mère par le nourrisson. *Enfance* (1948).

——Autoerotism. *The Psychoanalytic Study of the Child* (1949), vol. 3–4.

Spitz, R. A. and K. M. Wolf. *The Smiling Response: A Contribution to the Ontogenesis of Social Relations*. Gen. Psych. Monographs (1946), no. 34.

Wiesener, H. *Entwicklungsphysiologie des Kindes*. Berlin: 1964.

5. For an introduction to the question of bodily proportions in primates, there is important information in the following:

Babor, J. F. and Z. Frankenberger. "Studien zur Naturgeschichte des Gorillas." *Biologia Generalis,* vol. 6, no. 4.

Schultz, A. Studies on the Growth of Gorilla, etc., in: Memoirs. vol. 2. (Pittsburgh: Carnegie Museum, 1927).

——Fetal Growth and Development of the Rhesus Monkey. *Contrib. to Embryol.* (1937), vol. 26, no. 155 (with important comparisons between ape and man).

——Further information in *Primatologia.* Handbuch der Primatologie series cited in note 3.

6. Since Bolk's fetalization theory must be taken into account in all discussions of evolution, it is important to separate it from the distortions it has undergone through the use of irresponsible clichés. The following works present an introduction to this theory:

Bolk, L. *Das Problem der Menschwerdung*. Jena: 1926.

——Die Entstehung des Menschenkinns, *Verh. Akad. v. Wetensch. Amsterdam* (1924), vol. 23, 2d section.

Hilzheimer, M. Historisches und Kritisches zu Bolks Problem der Menschwerdung, *Anatom. Anzeiger* (1926–27), vol. 62.

Kappers, J. A. Orthogenesis and Progressive Appearance of Early Ontogenetic Form Relations. . . . *Acta biotheoretica* (Leiden: 1942), vol. 6.

Schindewolf, O. H. Das Problem der Menschwerdung, ein paläontologischer Lösungsversuch. *Jahrb. d. preuß. geolog. Landesanstalt* (1928), vol. 49, pt. 2.

7. The following table gives a few additional statistics on the relative birth weight in mammals; they should show above all that man does not represent an extreme in this regard, and that the "tolerable limits" for mammalian mothers are high enough to accommodate the birth of infants weighing much more.

	BODY WEIGHT (KG)		*Weight of Newborn in % of Adult Weight*
	Adult	*Newborn*	
PRIMATES:			
TARSIER	0.114	0.023	20.0
RHESUS MONKEY	6.2	0.435	6.7
GIBBON	6.2	0.37	5.9
ORANGUTAN	75.0	1.5	2.0
HUMAN	65.0	3.2	5.3
GORILLA	100.0	1.8	1.8
OTHER MAMMALS:			
BAT (GREAT BAT)	0.028	0.006	21.4
ANTEATER	28.0	1.4	5.0
PYGMY HIPPOPOTAMUS	225.0	7.0	3.1
HORSE	450.0	50.0	11.0
CATTLE	450.0	35.0	7.7
INDIAN ELEPHANT	2,500.0	90.0	3.6

8. Mangold-Wirz, K. Cerebralisation und Ontogenesemodus: bei Eutherien. *Acta anat.* (Basel: 1966), 63(4):449–508.

Portmann, A. *Cerebralisation und Ontogenese: Medizinische Grundlagenforschung* (Stuttgart: 1962) vol. 4; also in

Portmann, A. *Zoologie aus vier Jahrzehnten.* Munich: Piper, 1967.

Weber, R. Transitorische Verschlüsse von Fernsinnesorganen in der Embryonalperiode bei Amnioten *Rev. suisse Zool.,* (1950), vol. 57.

9. Lange, E. v. Die Gesetzmäßigkeiten im Längenwachstum des Menschen. *Jahrb. f. Kinderheilkunde* (1903), n.s., vol. 57.

Scammon, R. E. On the Time and Mode of Transition from the Fetal to the Postnatal Phase of Growth in Man. *Anat. Record* (1922), vol. 23, no. 1.

10. The interpretation by von de Snoo shows how ambiguous the conditions considered here are. He does not see the significant size of the child's head as the primary determinant for parturition, as the most widespread opinion assumes; to him, this large head is, more than anything else, an especially favorable factor in the maintaining of a stable position for the fetus within the pelvic area of a body whose torso is vertical. The crouching resting position, typical of all primates, creates the prerequisites for such a stable position for the baby. In favor of this positional relationship, de Snoo sees quite simply the positive factor, the selection factor, which, through the lengthy, continuous favoring of both the position of the baby's head and the vertical posture of the body, has ultimately brought about the attainment of erect locomotion. "Anthropogenesis would then be mainly an obstetrical question," de Snoo concluded, thus providing an example of the way such attempts at deduction oversimplify a problem. The situation to be explained is so complex that it is impossible to designate one or another of the collaborative factors as the decisive one, as the cause of all the other events.

De Snoo, K. *Das Problem der Menschwerdung im Lichte der vergleichenden Geburtshilfe.* Jena: Fischer, 1942.

Grosser, Otto. *Frühentwicklung, Eihautbildung, und Placentation der Entwicklungsgeschichte des Menschen.* Munich: J. F. Bergmann, 1927.

11. E. Gruenthal, *Fortschritte der Neurologie und Psychiatrie* (1936), vol. 8; J. Klaesi, *Schweizerische medizinische Wochenschrift* (1916).

12. Since the first appearance of these fragments (1944), the problem of a statistical interpretation of "cerebralization" has been worked out in particular detail at the zoology department of the University of Basel (see Portmann, *Cerebralisation,* in note 8).

Edinger, T. Paleoneurology versus Comparative Brain Anatomy. *Confinia neurol.* (1949), vol. 9, no. 1–2.

——Die Paläoneurologie am Beginn einer neuen Phase. *Experientia* (1950), vol. 6–7.
Portmann, A. Etudes sur la cérébralisation chez les oiseaux. *Alauda* (1946), vol. 14.
——Les Indices intra-cérébraux. *Alauda* (1947), vol. 15, no. 1.
——Cérébralisation et mode ontogénétique. *Alauda* (1947), vol. 15, no. 2.
Roessle, Robert and Frédéric Roulet. *Mass und Zahl in der Pathologie*. Berlin and Vienna: J. Springer, 1932.
Scholl, D. The Quantitative Investigation of the Vertebrate Brain and the Applicability of Allometric Formulae to its Study. *Proc. roy. Soc. B.* (1948), vol. 135.
Wirz, K. Zur quantitativen Bestimmung der Rangordnung bei Säugetieren. *Acta anat.* (Basel: 1950), vol. 9, no. 1–2; see Mangold-Wirz, in note 8.

13. The question we are suggesting here is described in detail in a study by G. Bally, *Vom Ursprung und von den Grenzen der Freiheit: Eine Deutung des Spiels bei Tier und Mensch,* (Basel: B. Schwabe & Co., 1945); 2d ed., 1966. On the problem of play, the following must be compared:
Allemann, C. *Über das Spiel*. Zurich: 1951.
Buytendijk, F.J.J. *Wesen und Sinn des Spiels*. Berlin: 1933; New York: Arno Press, 1976.
Inhelder, E. Zur Psychologie einiger Verhaltensweisen—besonders des Speils—von Zootieren. *Z. Tierpsychol.* (1956), vol. 12, no. 1.
Meyer-Holzapfel, M. Das Spiel bei Säugetieren, *Handb. d. Zool.* (Berlin: 1956), vol. 8, no. 2.

14. The following studies are essential to preliminary work of a new anthropology, and themselves lead to further examination of the subject:
Flitner, A. *Wege zur pädagogischen Anthropologie*. Heidelberg: Quelle & Meyer, 1963).
Gehlen, A. *Anthropologische Forschung*. Hamburg: Rowohlt, 1961.
——*Urmensch und Spätkultur: Philosophische Ergebnisse und Aussagen*. Bonn: Athänum-Verlag, 1956.
——*Der Mensch*. 4th ed. 1950.
Grene, M. *The Knower and the Known*. London: Faber & Faber, 1966.
Kunz, H. Über biologische Psychologie. *Schwiz. Z. Psychol.* (1949), vol. 8, no. 4.
Landmann, M. *Philosophische Anthropologie*. Berlin: Sammlung Göschen, 1955; *Philosophical Anthropology*. D. J. Parent, trans. Philadelphia: Westminster Press, 1974.

Plessner, H. *Die Stufen des Organischen und der Mensch*. 2d ed. Berlin: DeGruyter, 1965.
Storch, O. *Die Sonderstellung des Menschen in Lebensabspiel und Vererbung*. Vienna: Springer-Verlag, 1948.
——Erbmotorik und Erwerbmotorik. *Österr. Akad. d. Wiss.* (1948), vol. 1.
——Zoologische Grundlagen der Soziologie. *Österr. Z. öffentl. Recht.* (1950), vol. 3, no. 3.

15. See note 4; in addition, see G. Variot, *Bull. Soc. Anthropol.* (Paris) (1926), series 7, vol. 7.

16. Biologists who study language should pay particular attention to the humanities if they want to avoid a disastrous one-sidedness. The works cited here only indicate the path to follow toward a more comprehensive view of the problem.
Binswanger, L. Über Sprache und Denken. *Studia Philos.* (1946), vol. 6.
Delacroix, H. *Le Langage et la pensée*. 2d ed. Paris; F. Alcan, 1930.
Groos, K. Zum Problem der Tiersprache. *Z. Psychol.* (1935), vol. 134.
Kainz, F. *Psychologie der Sprache*. 2 vol. Stuttgart: F. Enke, 1941–43.
Lévi-Strauss, C. *La pensée sauvage*. Paris: Plon, 1962; *The Savage Mind*. Chicago: University of Chicago Press, 1966.
Revesz, G. *Ursprung und Vorgeschichte der Sprache*. Bern: A. Francke, 1946; *The Origins and Prehistory of Language*. J. Butler, trans. New York: Philosophical Library, 1956.
Ryffel, H. Die Sprache als Mittelpunkt der Bildung. *Schulpraxis* (Bern, 1949), vol. 10.
Stern, Cl. and W. *Die Kindersprache*. 4th ed. Leipzig: J. A. Barth, 1928.

17. An introduction to a discussion of "laughing" and "crying" can be found in the following works:
Buytendijk, F.J.J. Das erste Lächeln des Kindes. *Psyche.* (1947), vol. 2, no. 1.
Jünger, F. G. *Über das Komische*. Berlin: Widerstands-Verlag, 1936.
Plessner, H. *Lachen und Weinen*. Bern: Sammlung Dalp, 1950; *Laughing and Crying: A Study of the Limits of Human Behavior*. J. S. Churchill and M. Grene, trans. Evanston, Ill.: Northwestern University Press, 1970.
Stern, A. *Philosophie du rire et des pleurs*. Paris: Presses Univ. France, 1949.
These studies guide the reader to the significant older works on the subject.

18. The term "milk dentition" is to be understood in the human context, and is therefore difficult to use for mammals in general and often

should not be used. If a tooth that has no "replacement" can be regarded as a "milk tooth," then the three rear cheek teeth in humans are delayed elements of the first dentition, hence "milk teeth." On that account, in many studies, the third cheek tooth after the canine is even considered absolutely to be part of the milk dentition, bringing the total number of teeth in the first dentition to twenty-four. This must be taken into consideration when studies of mammalian second dentition are evaluated. We use the older arrangement of twenty teeth.

E. M. Lang, Austritt der Milchzähne beim Gorillakind. *Der Zoologische Garten* (1966), n.s., vol. 32, no.5.

19. For more precise information on transformation in Conrad, see note 20.

20. One concept that is close to my own was proposed by Professor de Rudder in Frankfurt a. M. back in 1937 (B. de Rudder, Allgemeinprobleme des Wachstumsalters, in R. Thiel, *Gegenwartsprobleme der Augenheilkunde*. (Leipzig: Thieme, 1937). I only came to know of it later and cite here the critical passages:

> If we take all these observations into consideration, the thought immediately occurs that the progressive urbanization of contemporary civilized countries is probably the critical factor. It is not just that today, one-fourth of the population of Germany lives in what are *undeniably* cities, one-third in cities determined to be such statistically, one-sixth in very large metropolitan areas (over five hundred thousand inhabitants), and one-eleventh in the huge metropolitan areas of greater Berlin and Hamburg-Altona (Hellpach).
>
> With these few figures in mind, make a rough comparison with the situation around 1900. Bear in mind that the absolute number of *large cities* has increased ninefold between 1871 and 1933. And furthermore, what was the *urban milieu* in cities around 1870 and how does it compare with contemporary situations?
>
> However, this urbanization, with its influences that in my opinion are critical for the issue at hand, has also spread deeper and deeper into rural areas, with the result that the *Landkind alten Stils* [old-fasioned country person] undoubtedly is on his way out and can be found today only in constantly shrinking, out-of-the-way areas.
>
> The contemporary *urban* milieu is an enormous complex: its complexity seems perhaps inexhaustible, particularly when its essential variety is compared with other ways of living. Do we expect that this fundamental change in milieu would leave no traces on the

physical type of the people who grow up in it? As for childhood development, it seems to me that critical influences of increased urbanization of milieu bring about in particular those countless psychological stimuli and impulses that, as *acceleration of development of intelligence,* indeed have the *urban* way of life at their core, but not as their only seat.

To give just one example, we are aware of the effect of shops, movies, radio, and the press (and do not forget the *Illustrierte,** which today is *studied* by almost all urban children eight to ten years old). We realize how traffic and learning to handle traffic accustom children as young as the first grade to move about safely in it and to use public transportation; they also learn to play their favorite ball games in the streets while constantly dividing their attention between the ball and the traffic—compare this with the carefree, unforced devotion to play seen in country children. We think of the *needs* awakened in modern children through advertising, for everything from toys to clothing; and let us not forget the progressive reduction in the number of siblings in the child's world, which often results in the *only child's* being drawn into the intellectual and typically urban problem-context of the adult environment unusually early; and the tempo of this problem-context itself has intensified over the last thirty or even fifty years.

I would like to remind further that a *sexual problem,* a *sexual curiosity,* hardly existed for the farm child, who was surrounded by the reproductive behavior of animals from his earliest years and took it for granted, as Springenschmid described so aptly in his magnificent study of the farm child *(das Bauernkind)*.

In short, we are aware of all the influences that lead to the speeding-up of intellectual development, to a mental *acceleration* that itself is to be regarded as a stimulus for the tempo of physical development. What formerly was *reserved* for children of the wealthy classes, the intellectual activity of Pfaundler's brain-damaged patients, is today, everywhere in civilized lands, increasingly the property of all, and has, additionally, increased in overall intensity. The *proteroplasia* of the well-to-do, which Pfaundler compared with the previously accepted delayed development in other milieux, has today become a common phenomenon.

In summary, we would also like to say the following: Urban

* A very popular, illustrated weekly newspaper.——Trans.

populations were always (even around 1900) ahead of rural populations in physical development (including sexual development). The urban proportion of the total population has increased considerably. Therefore, urbanization has affected a larger proportion of the total population. Furthermore, cities nowadays are much more urban than formerly, with the result that the earlier effects of an urban milieu must have become intensified. And finally, with all this comes the fact that in many respects, urbanization has already come to rural populations; the effects are already partially felt. This acceleration, or proteroplasia, through the pervasiveness of the urban mode of life casts its net wider and wider; the child of our time is more and more an urban child, whose biology still awaits a valid treatment—as a contribution, so to speak, to the scientific study of urbanization recently demanded so urgently by Hellpach (Pp. 6–8).

21. For an introduction to pubertal growth and its attendant problems:

Bennholdt-Thomsen, C. Die Entwicklungsbeschleunigung der Jugend. *Ergebn. inn. Med. Kinderheilk.* (1942), vol. 62.

Conrad, K. *Der Konstitutionstypus als genetisches Problem.* Berlin: Springer, 1941.

Grimm, H. Der gegenwärtige Verlauf der Pubertät bei der weiblichen Berufsschuljugend in Mitteldeutschland. *Zentralbl. Gynäko.* (1948), vol. 70, no. 1.

——Einige Individualbeobachtungen über Schilddrüsenfunktion, Wuchsform und Reifungsvorgang bei Mädchen. *Endokrinologie* (1950), vol. 27.

Hoeltker, G. Zur Frage nach dem Reifealter bei melanesischen und indonesischen Mädchen. *Acta Tropica* (1949), vol. 6.

Lommel, Fr. Betrachtungen über das menschliche Reifewachstum. *Jenaische Z. Naturwiss.* (1943), vol. 76.

Portmann, A. Die Entwicklungsbeschleunigung der Jugend als biologisches und soziales Problem. (Offprint from "Der Konflict der Generationen.") *Akademische Vorträge gehalten an der Universität Basel* (1966), vol. 4. (Contains recent literature.)

Schneider, E. Zur Frage der Wachstumssteigerung. *Z. menschl. Vererb.- u. Konstit.-Lehre.* (1950), vol. 29.

Seitz, L. *Wachstum, Geschlecht und Fortpflanzung.* Berlin: 1939.

Steinbeck, L. *Schilddrüse und Längenwachstum.* Wintherthur: Verlag Keller, 1956. (Dissertation originally.)

Tanner, J. M. Earlier Maturation in Man. *Scientific American* (1968), vol. 218, no. 1. (Cites additional literature from 1966.)
Wissler, H. Pubertät und Pubertätsstörungen. *Schweiz. med. Wochenschr.* (1943), vol. 73.
The bibliographies of these works lead to diverse older studies, which cannot be listed here.

22. R. Musil, *Über die Dummheit*. Schriftenreihe "Ausblicke". Vienna: Bermann-Fischer, 1937.

23. Bloch, S. Untersuchungen über Klimakterium und Menopause in Albino-Ratten: Third report—Histologische Beobachtungen am alternden Genitaltrakt. *Gynaecologia* (Basel: 1961), 152:414–24.

——Untersuchungen über Klimakterium und Menopause in Albino-Ratten: Fourth report—Das Altern des Uterusstromas. *Gynaecologia* (Paris: 1962), 154:196–304.
Bloch, S. and E. Flury. Untersuchungen über Klimakterium und Menopause in Albino-Ratten: First report. *Gynaecologia* (Basel: 1957), 143(4):255–63.
——Untersuchungen über Klimakterium und Menopause an Albino-Ratten: Second report. *Gynaecologia* (Basel: 1959), 147(5–6):414–38.
Comfort, A. The Life Span of Animals. *Scientific American* (1961), vol. 205, no. 2.
Dittrich, L. Beitrag zur Fortpflanzung und Jugendentwicklung des Indischen Elefanten in Gefangenschaft. *Der Zoologische Garten* (1937), 34:56–92.
Vischer, A. L. *Das Alter als Schicksal und Erfüllung*. Basel: B. Schwabe, 1942.
——*Seelische Wandlungen beim alternden Menschen*. Basel: B. Schwabe, 1949; *On Growing Old*. G. Onn, trans. Boston: Houghton Mifflin, 1967.
——Aufgaben und Begrenzung der Altersforschung. *Bull. schweiz. Akad. med. Wiss.* (1949), vol. 5.

On the biological problem:

Miescher, K. Zur Frage der Alternsforschung. *Experientia* (1955), vol. 11, no. 11.

Index